T0143383

Cognitive Radio – An Enabler for Internet of Things

RIVER PUBLISHERS SERIES IN COMMUNICATIONS

Indexing: All books published in this series are submitted to Thomson Reuters Book Citation Index (BkCI), CrossRef and to Google Scholar.

The "River Publishers Series in Communications" is a series of comprehensive academic and professional books which focus on communication and network systems. The series focuses on topics ranging from the theory and use of systems involving all terminals, computers, and information processors; wired and wireless networks; and network layouts, protocols, architectures, and implementations. Furthermore, developments toward new market demands in systems, products, and technologies such as personal communications services, multimedia systems, enterprise networks, and optical communications systems are also covered.

Books published in the series include research monographs, edited volumes, handbooks and textbooks. The books provide professionals, researchers, educators, and advanced students in the field with an invaluable insight into the latest research and developments.

Topics covered in the series include, but are by no means restricted to the following:

- Wireless Communications
- Networks
- Security
- Antennas & Propagation
- Microwaves
- Software Defined Radio

For a list of other books in this series, visit www.riverpublishers.com

Cognitive Radio – An Enabler for Internet of Things

S. Shanmugavel

Professor, National Engineering College
Kovilpatti, India

M. A. Bhagyaveni

Professor, College of Engineering, Guindy
Anna University, Chennai, India

R. Kalidoss

Professor, Sri Sivasubramaniya Nadar College of Engineering
Chennai, India

Published, sold and distributed by:
River Publishers
Alsbjergvej 10
9260 Gistrup
Denmark

River Publishers
Lange Geer 44
2611 PW Delft
The Netherlands

Tel.: +45369953197
www.riverpublishers.com

ISBN: 978-87-93519-40-4 (Hardback)
978-87-93519-39-8 (Ebook)

Contents

Preface

Communication technology is facing new challenges today from Internet of Things (IoT) model. These challenges arise due to the need to connect numerous heterogeneous objects. New radio technologies and network architectures require designs that can take care of several devices of the future having connectivity demands. Radio communications require exclusive adaptation by the frequency spectrum allocation to suit efficient spectrum utilization which takes into consideration new bandwidth and application requirements. Pursuit of innovative research directions based on the use of opportunistic radio resource utilization such as the ones based on Cognitive Radio (CR) technology is necessary. This is meant for efficiency and reliability.

Cognitive Radio is an enabler communication technology for IoT. The focus is currently on investigation of adaptation of the CR technology in objects of small size with the objective of minimizing wireless interference and also to ensure the concept of "always connected". The capabilities of cognitive radio have reached an advanced stage. They include features like spectrum sensing, knowledge of environment, learning and self adapting. These allow maintenance of efficient communication in a manner that means seizing of opportunities CR technology has the ability of automatic detection and learning from the radio environment. It can also adopt the transmission parameters in real-time share the wireless spectrum in a multi-dimensional manner, in space, time, frequency, modulation mode, which eventually improves spectrum efficiency. Cognitive Radio and Cognitive Radio Networking (CRN) are taken as key IoT enabling technologies. Their integration with future IoT architectures and services is expected to empower the IoT paradigm. With this in mind, we make a survey of Cognitive Radio approaches and challenges posed for M2M and IoT.

This book is designed for post graduate students in engineering and research scholars working in advanced wireless communication techniques.

Chapter Organization

The text comprises seven chapters: Chapter 1 introduces Internet of things and its importance in the next generation wireless networks. The enabler for IoT is cognitive radio. Software defined radio is a fundamental platform for development of CR. SDR and its architecture features are outlined in Chapter 2. Chapter 3 deals with cognitive radio and its types. The basic process involved in cognitive radio is spectrum sensing. Various types of spectrum sensing are discussed in Chapter 4.

The other important operations of cognitive radio are spectrum management, spectrum mobility and spectrum sharing, Chapter 5 outlined these topics. Upper layer and cross layer issues are also discussed in detail. The first standard for cognitive radio is IEEE 802.22 Wireless regional area network standard. Chapter 6 starts with the functional model of IEEE 802.22 followed by turnaround time and cross time slot issues in IEEE 802.22. The solutions for the shortcoming are explained with various simulation results.

Chapter 7 focus a various experimental operations in SDR and CR platform using MATLAB, WARP and USRP architecture model. The users will be able to develop and simulate their own algorithm using real time wireless model.

Acknowledgements

We are grateful to *Professor Huseyin Arslan of the University of South Florida*, who consented to review our book and the River Publication team for their valuable insights and suggestions, which have greatly contributed to significant improvements of this text.

List of Figures

List of Tables

List of Abbreviations

3GPP	third Generation Partnership Project
ADC	analog to digital converters
aMS	Adjacent Mobile Station
AWGN	additive white Gaussian model
BD	Bayesian detector
BS	base station
CCC	Common Control Channel
CDF	Cumulative distribution function
COBRA	Common Object Request Broker Architecture
CogNets	Cognitive networks
COTS	commercial off-the-shelf
CPE	consumer premise equipments
CR	Cognitive Radio
CRN	Cognitive Radio Network
CSCC	common spectrum coordination channel
CSMA	carrier sense multiple accessing
DAC	digital-to-analog converters
DC-MAC	decentralized cognitive MAC
DDC	Digital Down Converter
dMS	Desired Mobile Station
D-QDCR	QoS based dynamic channel reservation
DS	downstream
DSAN	dynamic spectrum access Networks
DSL	digital subscriber loop
DSP	digital signal processing
DUC	Digital Up Converter
ED	energy detector
FCC	Federal Communications Commission
FDD	Frequency Division Duplex
GLRT	Generalized likelihood ratio test
IERC	European Research Cluster on the Internet of Things

INFOSEC	information security
IoT	Internet of Things
ISR	ideal software radio
JTRS	*Joint Tactical Radio System*
KL	Kull back-Leibler
LAN	local area network
LOS	line-of-sight
LRT	likelihood ratio test
M2M	Machine to Machine
MAC	medium access control
MLE	maximum likelihood estimates
NFC	Near- Field Communication
NP	Neyman-Pearson
OMG	Object Management Group
PD	performance of detector
PD	probability of detection
PDF	probability density functions
PDR	programmable digital radios
PFA	probability of false alarm
P-MP	point to multipoint
PU	*primary users*
QoS	quality of service
RFID	Radio Frequency Identification
RKRL	Radio Knowledge Representation Language
RTO	retransmission timeout
RTT	round trip time
SCA	Software Communications Architecture
SCH	super frame control header
SDH	synchronous digital hierarchy
SDR	Software Defined Radio
SNR	signal-to-noise ratio
SU	Secondary users
TDD	Time Division Duplex
TRANSEC	transmission security
TS	test sample
TVWS	TV whitespaces
UCS	Urgent Coexistence Situation
UHD	USRP Hardware Drive
UP	upstream

URI	Uniform Resource Identifier
USRP	Universal Software Radio Peripheral
WARP	Wireless open Access Research Platform
WISP	Wireless Internet Service Providers
WRAN	wireless regional area network
WSN	wireless sensor networks
xG	Next generation

1

Introduction

On some future date, strange, and incredible things may happen. Curtains are up on the instruction of our alarm clock. Other breath taking things that happen at the initiative of the alarm clock are: the coffee maker knows when to make coffee; the car melts the snow that has gathered overnight. Other things that happen are our refrigerator texts our grocery lists, our washing machine announces the completion of laundering, and our doctors update their prescriptions with the use of data beamed from tiny sensors attached to our bodies. These are not figments of imagination or just wishful thinking or fantasies. We can bet on these happening, becoming a reality, through Internet of Things (IoT) which is certain to enable communication with each other, developing their own intelligence. Cisco systems have predicted that, by 2020, the Internet will have more than 50 billion connected things including televisions, cars, kitchen appliances, surveillance cameras, smart phones, utility meters, cardiac monitors, thermostats and almost everything that touches our imagination. Cisco projects a profit of US $14.4 billion for the IoT technology globally in the next decade, with IoT connecting our things of everyday use (e.g., keys, refrigerators, and washing machines), pets, energy grids, health care facilities, transportation systems etc., to the Internet. This is illustrated in Figure 1.1.

1.1 Features and Application of IoT

The IoT has caught up the attention of common people and researchers alike, conjuring up the vision of connecting many things in this world. This will surely transform our ability to interact with real world objects, process information, and enable our decision making capability with savings in time and money. IoT means the next phase of internet wherein everything will be seen. The number of objects that get connected to the internet is growing exponentially day by day. The penetration of connected things is expected to

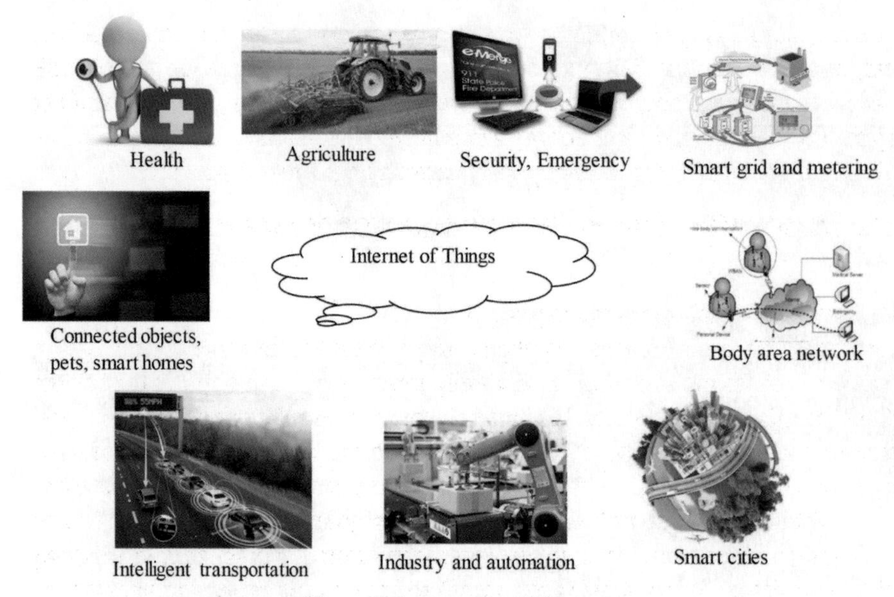

Figure 1.1 Internet of Things.

reach 2.7% of the total by 2020 from 0.6% in 2012. There are forecasts that, by 2020, everyone will get connected to at least two things. The emerging M2M (Machine to Machine) technology will eventually accelerate the growth of IoT technologies which connect machines, devices and objects to the internet, transforming them into "intelligent" assets. 11.6 billion-this is the number of mobile connected devices by 2020, including M2M modules. This will outstrip the world's population which is extrapolated as 7.8 billion. The global projection of M2M connections is 604 millions in 2015 going to a staggering 3.1 billion by 2020. M2M traffic flows would form a big slice of the entire Internet traffic in the coming years. Manufacturing and health care will be the largest IoT market segments according to market specific surveys. Oil and gas will be leading sub-segment of manufacture together with the energy section, mobility, and transportation. In the sphere of consumer applications, home automation items (smart thermostats, security systems and refrigerators) will dominate the market in the next few years.

A good momentum on IoT standardization and IoT workshops is witnessed today. This facilitates interoperability testing events which help reaching consensus on the development of IoT standards. IoT is a popular expression in the computing industry. It is seen in the marketing campaigns of major networking companies like Cisco and microprocessor giants like Intel. It has also figured in International Conferences like "Internet of Things" world forum with a view to ensure successful IoT deployment, IETF working groups like 6Lo, ROLL, 6TiSCH are in the process of developing standards that can enable seamless integration of low-power wireless networks into the internet through solutions for address segment and routing. Simultaneously, the third Generation Partnership Project (3GPP) is on the move for providing support to M2M applications on 4G mobile networks like UMTS, LTE/LTE-A. The objective is to facilitate embedding of M2M communications in the future 5G systems.

The application of IoT products will be all pervasive and will manifest all perceivable areas ranging from smart phones and going up to smart cities. Education, manufacture, healthcare, mining, oil and gas, energy, commerce, transportation, surveillance, infrastructure, supply chain, logistics- the segment or area-IoT products will be there. Opportunities for the use of IoT will be infinite. Its realization will manifest in more and more devices coming within the fold of IoT. This phenomenal development will owe it to the availability of tiny, low power and low cost devices, as for example, sensors, actuators, RFID tags, low power tiny computers. The world will also witness advanced technologies and protocols, as for example, M2M, WSN, RFID, Cognitive radio, IPV6, GLOWPAN).

The IPv6 protocol is the new version of IP with a large address space (2^{128} addresses). It would have the ability to provide identification/address for almost all objects worldwide. M2M communication and other similar featured technologies will help implementation and deployment of IoT applications. Efficiency and reliability will be the features with wireless networks and wireless sensor networks (WSNs) that can enable communication of data to and fro from "things". Potential users will also be able to develop their own application software due to the availability of a large number of development platforms. Interoperate protocol stacks and open standards for practical realization of IoT will also benefit by the standardized bodies.

1.2 Enabling Technologies and Protocols for IoT

An enormous potential for diverse applications in day to day life through IoT is clearly seen. Advanced technologies and protocols that have emerged recently have provided support to this. An overview of these has been the subject matter of a recent survey. This section seeks to make a review of the existing key technologies that include RFID and WSN. These technologies support not only the provision of IoT services but also the M2M technologies that have come up enabling direct connectivity to the "machines" to the internet, transforming them into "intelligent" assets with ability to communicate. This section also provides a brief discussion on the schemes involving naming and addressing for IoT and also the techniques for data storage and analysis supporting the IoT vision.

1.2.1 RFID and Near-Field Communication (NFC)

Radio Frequency Identification (RFID) is expected to provide a solution to ensure availability of M2M and IoT communication for everyone, with focus on economy, keeping the cost low. Tags fixed to or incorporated into physical objects to enable their identification help realization of RFID. Each RFID tag comprises a microchip attached to an antenna. It can be read through radio waves in almost every spectrum band despite RFID using unlicensed bands mostly. RFID enables wireless access to data (e.g., identification numbers) stored on microchips, using radio waves. This way, RFID tags do the role of electronics bar codes that enable automatic identification of things to which they are attached. Currently, RFID tags are finding use in retail and supply chain management, access control, transportation and electronic passports. With the advent of the recent advances in RFID technology, cheaper tags for monitoring applications in IoT (e.g., for monitoring temperature and location) are available today. The size of RFID systems is growing enormously to thousands of tags for various applications. Applications include transmission collusions, such as tag-to-tag collisions, reader-to-tag collisions, and reader-to-reader collisions which may occur in situations where there are many readers and tags in close vicinity. Anti-collision protocols provide solutions to problems arising out of collision helping improvement in performance. RFID is considered as a key technology which can accelerate the formation of IoT. The most recent use of RFID technology is seen in the form of near-field communication (NFC) supporting electronic payments. RFID standard provides the base or foundation on which NFC technology is built. NFC is a short-range communication standard wherein devices have the ability to engage in radio communication among themselves when in contact or brought into close proximity to one another. With the introduction of smartphones that include NFC transceivers, phones can read passive NFC tags which can store a URI (Uniform Resource Identifier) while still being small, thin, inexpensive and attachable to almost anything. A matter of interest is the Apple's recent introduction of NFC support in its iphone6 – a device that facilitates wireless payments using Apple Pay.

Apple has also introduced NFC chips in the Apple Watch. Besides, there are plans to continue the use of NFC in devices that would see introduction at a future date. This will glorify NFC into becoming a key IoT enabler there by proving to be a remarkable health in Apple's cap.

Wireless sensor network (WSN) has some unique features like being autonomous and easily deployable. These have provided WSN the status of being another key enabling technology for IoT deployment. Sensor networks can cooperate with the RFID system and enable efficient tracking and collection of information on position, movement, temperature etc., In addition, technological advances of recent origin in low power integrated circuits and wireless communications have helped in getting efficient, low cost, low power miniature devices which can be used in remote sensing applications. As a result, there is the possibility of collection, processing, analysis and dissemination of information gathered in a wide variety of environments in IoT through sensor networks that consist of a large number of smart sensors. As against this, there are a few limitations that WSN suffers from. These include resource constraints in terms of computing and power, with the requirement of appropriate action, to ensure effective use of WSN technology. In the backdrop of IoT, research issues in WSN that require study include low cost routing discovery and energy conservation.

Untuned radios can achieve a throughput comparable to a tuned network. This is possible through the use of network coding in WSNs. Some examples of employment of the network coding scheme in WSNs are those for dealing with the multipath problem in sensor networks and for efficient data dissemination, storage and collection in WSNs. However, this requires scalability in the network coding algorithm and also stability for a large scale WSN deployment. There should also be consideration of a dynamic network topology and routing features of WSN. Complexity of intermediate nodes in a resource limited WSN network is seen through network coding. It also results in additional transmission delay which is rather inadvisable for real time applications. Hence, there is need for further study for network coding in WSN and IoT.

1.2.2 M2M Technologies

M2M technologies of recent origin act as a key enabling technology for the pragmatic realization of IoT. They provide connectivity between devices all over along with the ability of autonomous communication without human intervention. With IoT in the background, addition of the human element to machine interaction makes the scope of M2M applications wider.

M2M communications have the potential for transforming the day to day life through enabling of smart homes, smart grids, smart transportation, smart buildings and smart cities. M2M connections are known to be an integral part of IoT vision.

On the whole, M2M communications are expected to grow to 3.4 billion globally by 2024. But, several issues require investigation for the growth of M2M-IoT. One such issue is the fragmentation created by multiple proprietary M2M solutions which do not have interoperability with other M2M systems or solutions. One M2M is the worldwide initiative for interoperability of M2M and IoT devices and applications. Its objective is to develop specifications for a common M2M service layer platform which builds on the existing IoT and web standards, defining at the same time, specifications for protocols and service API's. Another M2M specification provides a framework for supporting applications and services like smart grid, connected car, home automation, public safety and health. A third M2M is known for its recent publication of its Release 1 standards which provide specifications for enabling optimized M2M interworking and developing a platform for M2M and IoT devices and applications. Recent advances and the spread of M2M technologies have helped development of several M2M platforms, like, for instance, the open MTC platform which is a prototype implementation of the M2M middleware which has provision for a standard-compliant platform for developing M2M and IoT applications as its objective. One more M2M defines the specification for interoperability of IoT platforms at the service layer. But, many implementation details are open, despite the provision of details of functionality, protocols and API's of platforms. Among such details are lack of scalability, availability and deployment aspects of the IoT platforms that implement these services. There is also another issue which is the limited availability of the wireless spectrum as M2M expansion requires additional spectrum. The number of M2M devices is constantly on the increase. This has triggered a large number of connected devices, visualized for the IoT, creating a major challenge manifested by spectrum capacity. The TV whitespaces (TVWS) hand down a solution for M2M, simultaneously fitting well with the operation of M2M nodes with features of low speed and low duty-cycle.

1.2.3 Naming and Addressing Schemes for IoT

Dealing with the problems involved in M2M and IoT is very important as their success relies very much on the ability to identify devices/things in a unique manner. Identification of all connected devices is imperative and this can be accomplished by their unique identification, location and functionalities.

This enables exclusive identification of billions of devices and, additionally, control remote devices through the internet. Any exclusive scheme dealing with the problems should have features like uniqueness, reliability, persistence and scalability. IoT has varied network technologies and system architectures which make it a heterogeneous paradigm. It is this that makes integration of the different types of networks an important issue in IoT. The IP protocol provides an empirical solution (in the form of IP based network) for communication technology for the wireless network technologies in vogue including WiFi, IEEE 802.15.4, and RFID.

The new version of IP i.e IPv6 has a larger address space and mobility feature. It enables interconnection of everything and every network node. It also provides a solution for communication between different systems featuring a global unified address allocation and routing. However, investigation of several issues such as IPv6 header compression, mobility, security and QoS is required for the use of IPv6 addressing in the backdrop of IoT. IPv6 should consider QoS and real time requirements for IoT real time applications.

1.2.4 Data Storage and Analysis Techniques

The IoT paradigm has the feature of generating voluminous data, with need for innovative methods for data storage and analysis and energy conservation. With the rise in energy consumption, there is need for data centres to replace new ways of energy conservation, for example, data centres that run on harvested energy and are centralized, providing energy efficiency and reliability. Storage and intelligent use of data are imperative for smart monitoring and actuation in Iot. This in turn requires the development of artificial intelligence algorithms (centralized or distributed) and fusion algorithms to enable analysis of the gathered data. More than this, the most updated machine learning methods, genetic algorithms, neural networks, and other artificial intelligence techniques are imperative for effective automated decision making. As a general rule, novel intelligent systems with features like interoperability, integration and adaptive communications are appropriate for IoT communications coming up. A centralized infrastructure for supporting storage and analysis is also important for IoT.

1.2.5 Cognitive Radio for IoT and M2M

The prevailing and emerging technologies include Wi-Fi/IEEE 802.11, HSDPA, LTE/LTE-A, 5G, ZigBee, Z-Wave, Bluetooth Low Energy 4.0 and other IEEE 802.15.4 standards. These are known to implement IoT applications, and operate in non-licence ISM bands that are getting congested. This phenomenon has given rise to new challenges that have the objective of dealing with the management and usage of spectrum resources for efficient and effective utilization of IoT, which holds at promise of being a part of the future interest, covering almost the entire gamut of domains, industry and sector. It can generate enormous wireless access data including M2M communication as also machine to human communication. Any indifference to this problem will result in the emergence of a situation witnessing paucity of spectrum recourses that will become a bottleneck for the development of IoT in the future. It will mean a high priority to ensure availability of adequate spectrum for handling the traffic involving billions of new wireless nodes that are getting connected to the internet. CR technology will provide benefit to IoT by enabling additional efficient radio spectrum usage. IoT is becoming a reality. This would indicate continuously increasing demands for spectral resources that would lead to overcrowding of ISM bands. At the same time, there are indications of spectrum utilization measurements in many unused or underused licensed bands over space and time, as for example, spectrum bands for TV broadcasting resulting in huge spectrum wastage.

This has triggered regulatory agencies such as Federal Communications Commission (FCC) for opening licensed bands to unlicensed/secondary users through use of CR. It is well know that CR, as an improving technology, has the capability of improving spectrum usage and reducing spectrum scarcity which are exposed to the hazard of exploiting underused spectral resources through reuse of the unused spectrum as an opportunity. Traversing similar lines, IEEE has been the emerge for IEEE 802.22 WRAN working group for development of a standard for a Cognitive radio based network for secondary access in TV white space (TVWS) known for their non-interfering nature and opportunistic attitude. The research community has considered CR for the management of dynamic radio spectrum as, for example, in ISM bands and as secondary users in unused TV bands. In addition, several leading organizations like Motorola, Philips, Qualcomm are making huge investments for development of CR technology of late, the European Research Cluster on the Internet of Things (IERC) has provided a duration for the standardization of IoT technologies. An IERC project, viz., RERUM is on the job of adapting CR on the IoT devices.

1.3 Requirement and Challenges of IoT

1.3.1 Heterogeneity Issues

IoT applications are in a wide range on the basis of design, deployment, mobility cost, size, resources, energy, heterogeneity, communication modality, infrastructure, network topology, coverage, connectivity, network size, lifetime, etc., they introduce complexities in the development of IoT applications in ways more than one. An amalgam of software and hardware platform is not adequate for offering support to the entire design space. This brings in the need for the use of heterogeneous systems. Inter operability and heterogeneity related issues are causing great concern. There is also the problem of having to manage the devices of different hues including unique characteristics of a wide variety. In addition to this, there is the presence of heterogeneous technologies of various shades connecting the objects to the network. Such technologies have given rise to the need for exploring synergies with potential for dealing with functionalities like intelligent connectivity for smart objects. New models of communication technologies are needed for dealing with heterogeneity issues, technologies that can provide environment discovery, self-organization, and self-management capabilities.

1.3.2 Flexible, Dynamic and Efficient Networking and Communication

Many IoT applications need higher QoS in combination with adaptability and re-configurability features. But, there are of a large number of heterogeneous sensing devices of small size used in such applications. Management of such devices is not possible through use of conventional tools. There is also the importance of plug and play objects with self-discovery, self-configuration, and automatic software deployments. Such importance is from the end user's point of view. Networking should have the capability to support multi-hops. It should be dynamic and efficient too. Radio communications need adaptation of frequency spectrum allocation for efficient spectrum usage taking into account new bandwidth and application requirements. A static spectrum allocation for different applications does not ensure efficiency.

1.3.3 Self-Organization, Re-Configurability and Automaticity

Specific ability and resources are necessary for smart objects for their reconfiguration. A transparent insertion of objects in the environment without any external action is needed for reconfigurability. Ability to scan the environment, detection of neighbouring objects and sub reconfiguration are the essential requisites for objects. There is also the ability of nodes to play different roles in accordance with situation and capabilities, as for example, as a relay or leader that can control a sub network or collect and steer information. A significant challenge is in the area of implementation of reconfigurability and flexibility, with a low energy budget at the same time.

1.3.4 Energy Efficiency

IoT has objects of a wide variety including low power sensors; smart phones etc., with variegated power consumption profiles. Energy efficient designs and communication need dealing with for a longer duration of IoT objects as also for a "green" design. Energy management should have the adaptability to different ways of opportunistic energy harvesting and scavenging. The need for energy communication arises due to the communication part being the main source of power consumption. There is also the requirement for a communicating object to have adequate processing power to ensure efficient energy management. This leads to the need for designing energy optimization and management algorithms to consider the low processing capabilities seen in some of the IoT devices.

1.3.5 Cooperative and Ambient Intelligence

Intelligence tools with capability of ambient and cooperative intelligence are used in IoT. There is also the need for intelligent distributed agents which can perform many management functions in a decentralized fashion, with local dealing with simple situations with good response. Granting the presence of cooperation and communication among neighbors, smart objects can enhance their knowledge of local environment and make coordinated decisions within the ambit of a network. Smart interfacing between the digital and the real world needs intelligence for a system with ability to sense the real world, to communicate information, to process raw information into a meaningful one and act on it too.

Evolution of CR involves three main steps. The first step is the software defined radio (SDR) that provides support to its implementation. Figure 1.2 explains the evolution of CR from software defined radio to cognitive radio. This cognitive concept can be extended to Cognitive networks

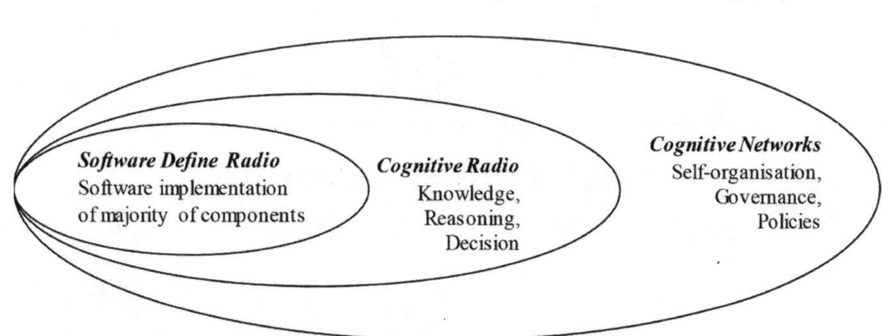

Figure 1.2 Evolution of Cognitive Radio.

(CogNets), which comprise three parts viz., SDR, CR and CogNets with the feature of highlighting the capabilities provided by each. The upcoming Chapter 2 describes the fundamental concepts associated with SDR and its architecture.

2

Software Defined Radio

2.1 Introduction

Following the incorporation of the times software was incorporated into radio and radio technologies, the main goal of radio developments has been software defined radio or software radio. The complete configuration of radio using software, ability to change the configuration at a particular instant to enable the use of software as a common access platform have been the key objectives of SDR. Re-configuration of radio using software depending on the scope of operation, change of roles or up-gradation of radio have been the subject matter for research for many years past.

The Joint Tactical Radio System is one such system in which one waveform from a plethora of waveforms can be selected and communication carried out. Feasibility of this system becomes an accomplishment through use of a common hardware platform by simple reconfiguration and reload of the radio. The distinction and popularity of this particular application in military can be attributed to its capacity to ensure communication between troops from different countries in a common platform i.e., coalition style operations.

On the commercial front, the SDR can find application in cellular networks, subject to frequent changes in standards. This facilitates smooth transition from one standard to another through reconfiguring the system using software instant any contact with the hardware platform. Migrations from one standard say UMTS to HSPA, another standard and further to a higher one say LTE can be done by just loading the software and reconfiguring it to suit the needs of the user, despite the presence of divergent standards in the modulation techniques. Thus the SDR has a high potential in many fields and its use is bound to increase with the developments in the field of communication.

2.2 Definition of SDR

The definition as given by the SDR forum is as follows:

2.2.1 Software Controlled Radio

The radio in which some or all of the physical layer functions are Software Controlled, in other words is a technique where the functions of a radio are prescribed by the software.

2.2.2 Software Defined Radio

Radio in which some or all of the physical layer functions are Software defined. In other words, it is the usage of software to determine the functions and the specifications of a radio. Thus changes find reflection in the radio also, when the software is subject to any modification.

An all encompassing definition that captures the essence of SDR is that it is a technique that comprises of a generic hardware platform in which software that carries out functions like modulation, demodulation, filtering, frequency selection and frequency hopping (if needed), operates. This demonstrates the performance changes in the radio arising from the reconfiguration of the software. The software employs digital signal processing (DSP) processors as well as general purpose processors to carry out the required functions.

The ideal emanation of signals from the transmitter and the processing at the receiver side after reception and conversion of the signals to bits would require a digital to analog convertor at the transmitter side to deliver very high power and extremely minimal noise. This reveals the impediment to a complete definition and control by software through a few limitations. It is shown in Figure 2.1.

2.3 Levels of SDR

Since the definition and control of radio using software is dependent on various parameters, the SDR Forum (now called the Wireless Innovation Forum-WINNF), has defined various tiers, providing gives a broad perspective of the tier for a particular radio system to fit in.

Tier 0:- This represents a type of radio that cannot be controlled by software i.e., non-configurable hardware.

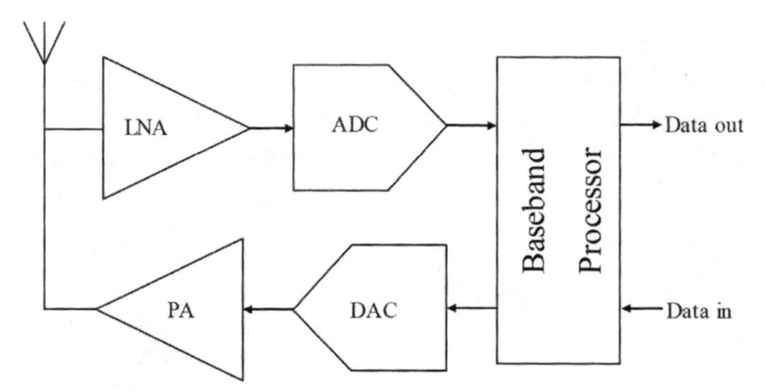

Figure 2.1 Block diagram of an 'Ideal' software defined radio.

Tier 1:- Represents a radio system in which only limited functions like power level, interconnections can be controlled by software, however, no control over mode and frequency.

Tier 2:- This represents a system in which a significant number of parameters like frequency, modulation and waveform generation/detection, wide/narrow band operation, security etc can be controlled by software. However the RF front end remains hardware based and is non-reconfigurable.

Tier 3:- An ideal software radio (ISR) in which the "front end "is also configurable. This is considered to have complete programmability with the boundary between configurable and reconfigurable elements lying very close to the antenna.

Tier 4:- This is one step ahead of the tier 3 system where it supports multiple functions and frequencies in addition to complete programmability. This is also termed as the Ultimate Software Radio (USR). With mobile communication developing by leaps and bounds, this technology finds application in many devices that include a software definable multifunctional cell phone.

2.4 SDR Waveform Portability

Waveform portability is another major advantage of SDR. The factors that explain the need for waveform portability and explained below:

Cost savings: The cost of usage of different waveforms for communication in military and commercial levels is very high and hence there is an indispensible need for techniques that facilitate the reuse of waveforms for communication.

Obsolescence mitigation: This is similar to cost savings in areas where there is a need to transfer the existing waveforms into newer platforms due to the advent of new hardware techniques.

Interoperability: There are many requests from customers to have the ability to use the systems across platforms on the basis of complete interoperability.

Even though waveform portability is not easy to achieve, it is necessary to introduce measures for realizing this in the early stages of design and development. This vouches for optimization in portability. Additionally a number of tools and elements like the SCA – Software Communications Architecture, and CORBA, a form of middleware associated with SCA come in handy for design purposes. There is also the need for many short-cuts and well-structured programming techniques for designing purposes. This is due to the persistent need to compile the codes in different platforms and hence programmed software that can be run on different platforms is required.

2.5 SDR Security

A high degree of security is expected in many communications seen military and commercial applications. In addition to the security demanded by the user, another security requisite in the case of SDR is secure software upgradation requirement that calls for a high usage of the internet. The high usage of internet will call for its use in software upgradation in SDR. This might give room for malicious software upgrades and other impediments in operation of the radio due to the corruption of software. Thus the security aspect of SDR should be seriously considered.

2.5.1 SDR Interoperability Testing

Interoperability testing ensures the ability of a code for waveform portability from one platform to another, providing the necessary functionality in the new platform. Realization of this needs employment of certified and accredited waveforms.

The major limitations that prevent the usage of SDR in many real time applications are the high processing power and the power consumption of the SDR. Since the decision to have a trade off in power consumption/processing power should be done even at the initial stage, SDR is not suited for cell phone designs. But cellular networks base stations employ SDR since the issues of power consumption and space constraints do not affect them much.

Further upgradation can also be handled efficiently by the base stations. It is also used in military and many other devices using SDR have started entering the market. But there is always the dilemma of choosing SDR as a technique for mobile communication. It is a fact that the implementation of SDR for small, cheap radios with lesser scope of change is a waste of resource. But SDR is an extremely efficient technique that should be considered for complicated systems with high likelihood of change and a good duration of service required.

2.5.2 SDR Hardware

An important aspect of the SDR technique is the hardware on which the software has to be programmed. The hardware is a fundamental requirement for SDR that serves as a platform on which the SDR runs. The design of a flexible hardware is, therefore, an important and interesting part of the design process. Though the entire technique revolves around the employment of software for operation, the hardware is the one that defines the ability of the software. In terms of design, performance and cost the challenges arise due to the requirement of the presence of the interface between software and hardware very close to the antenna to ensure higher levels of software control and reconfiguring ability. This calls for many decisions at an early stage regarding the cost, boundary locations based on the required functionality and performance expected.

The basic functional blocks are:

1. ***RF Amplification***: This is used for amplification of signals coming from and travelling to the antenna. There is a great need for a RF amplification block arising from the low probability of a DAC at the transmitter side. Amplification of the signals is extremely important in the receiver side as well considering the emergence of the problem of quantization noise when the signals are passed unamplified into the antenna, despite all the frequency specifications being within limit.
2. ***Frequency conversion***: The design SDR may or may not include certain analog signal manipulation section that indulges in frequency conversions. These analog sections carry out conversions to and from the final frequency or alternatively some intermediate frequency processing may also be carried out.
3. ***Digital Conversion***: The pivotal stage in which the conversion between digital and analog formats are carried out. The parameters that are to be considered in the transmitter side while carrying out this conversion are the maximum frequency and the required power level while the

parameters like maximum frequency and the number of bits to be assigned to a particular quantization level are to be considered in the receiver side.

4. ***Baseband processor***: A number of functions like digital conversion of the incoming and outgoing signals in frequency are carried out by the central baseband processor. Digital Up Converter (DUC) is the name assigned to these processors, since the conversion of signals from the base frequency levels to the required output levels is carried out by these elements. Similarly, the receiver side, the exact opposite process carried out in the transmitter side is done which involves bringing the signal down in frequency. This is done by a Digital down Converter (DDC) after which the signal needs filtration and demodulation and extraction of data.

Power consumption is a major issue in baseband processor. The processing level is directly proportional to the current consumption which, on increase calls for a higher level of cooling. The formats of the processors like general processors, DSPs, ASICs and particularly FPGAs also influence power consumption. There is considerable importance given to FPGA based design due to its capability of being reconfigured to the required specifications.

2.6 Software Radio Functional Architecture

The essential communication functions viz., source coding and channel coding face the need for expansion. The needs can be met through advances in technology that have done and, are still doing, the mighty role of ushering in new radio capabilities. A software radio/model has captured these new functional aspects.

2.6.1 The Software Radio Model

As a first step, multiband technology accesses simultaneously more than one RF band communications channel, which is then generalized to the channel set. This is shown in Figure 2.2. While this set includes RF channels, radio nodes like the base stations of the PC's and portable military nodes are also interconnected to a fibre cable, explaining the reason for their inclusion in the channel set.

The channel encoder of a multi band radio includes RF channel access, IF processing and a modem. Wideband antennas and multi-element arrays of smart antennas are included in the RF/channel access. This segment is known to provide multiple signal paths and RF conversion that span multiple

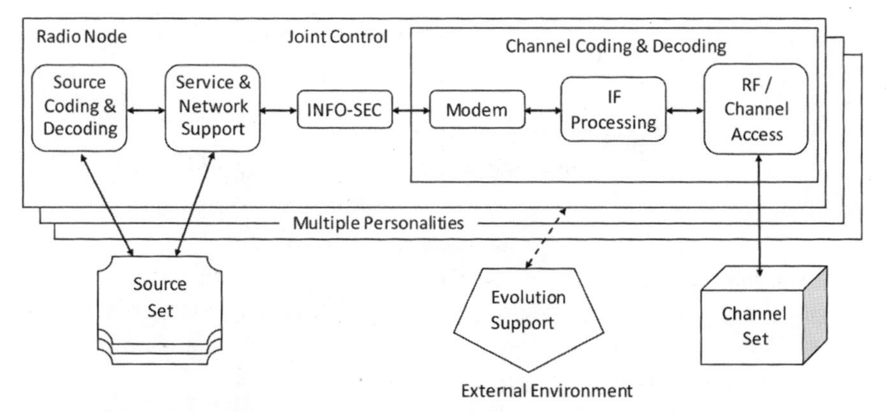

Figure 2.2 Functional model of a software radio communication system.

RF bands. IF processing is likely to include filtering, further frequency translation, space/time delivery processing, beam forming and related functions.

Multimode radios generate multiple air interface wave forms (modes) defined basically in the modem, the RF channel modulator-demodulator. These waveforms may be in different bands and may even span multiple bands. A software defined personality has, in its fold, RF band, a channel set (e.g. control and traffic channels), air interface waveforms and related functions. Despite many applications not requiring information security (INFOSEC), incentives for its use still exist.

Authentication reduces chances of fraud, while stream encipherment ensures privacy. Both have the specific advantage of ensuring data integrity. The fact of a communication event as, for example, through use of spread spectrum technique is not discussed by transmission security (TRANSEC). This is the reason why INFOSEC finds inclusion in the functional model despite the function being null in respect of many applications.

As an additional feature, the source coder/decoder pair currently includes the data, facsimile, video and multimedia source found essential for new services. Some sources are physically remote from the radio node, with connection via the synchronous digital hierarchy (SDH), a local area network (LAN) etc., through service and network support (detailed in Figure 2.2). Implementation of these functions may be done in multithreaded multi processor software orchestrated by a joint control function, which ensures system ability, error recovery, timely data flow and isochronous streaming of voice and video. With advancement in radios, increasing complexity is seen in joint control with evolution towards autonomous selection of band, mode and data format. Any of the functions has the likelihood of having singleton as, for example, single band versus multi bands leading to further complication of joint control. Additional users find support from an agile beam forming which enhances the quality of service. (QoS). The beam forming currently requires dedicated processor. However, these algorithms may, in future, time share a DSP pool along with the Rake receiver and other modem functions.

A point to note is that joint source and channel coding also yield waveforms that are computationally intensive. Introduction of large variations into demand through dynamic selection of land, mode and diversity as functions of QoS is seen. Such variations cause conflicts for processing resources. A further complication in the statistical structure of the computational demand is created by channel strapping, adaptive waveform selection and other forms of data rate agility.

There is also the loss of processing resource through equipment failures. Joint control gives the facility of integration of fault nodes, personalities and support functions on processing resources which include ASICs, FPGAs, DSPs and general purpose computers for ensuring the object of telecommunications.

It is possible for the user to upload a variety of new air interface personalities in a software radio. These may modify any aspect of the air interface under any circumstances which could be the waveform hoped, spread or otherwise constructed. The resources required (e.g. RF access, digitalized bandwidth, memory and processing capacity) should not exceed those available on the radio platform. There is, therefore, the need for a compelling need for mechanism of some kind for evolution support to define the waveform personalities, download them (e.g. over the air) and to ensure safety for each new personality prior to getting activated. It is necessary for the evolution support function to include a software factory. However, there is the requirement of support to the evolution of the radio platform both in the analog and the digital hardware of the radio node. Accomplishment of this is possible via the design of advanced hardware modules in an integrated evolution support environment or through the acquisition of commercial off-the-shelf (COTS) hardware modules or both. Figure 2.2 is the block diagram of the radio functional model amounting to partitioning of the black box functions of the ideal software radio nodes that have been referred to above with specific functional components.

2.7 Classes of Software Defined Radio (SDR)

The software radio parameter space illustrated in Figure 2.3 represents radio implementation as a function of digital access bandwidth and programmability. The horizontal access has programmability as the ease of making changes.

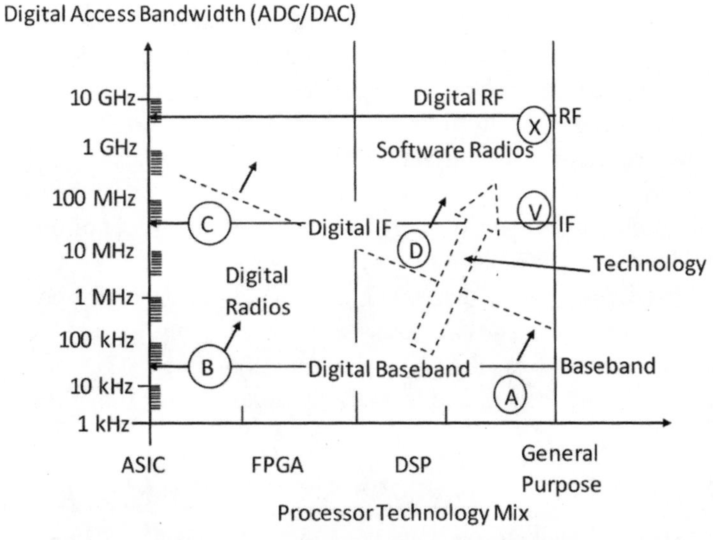

Figure 2.3 Software radio parameter space.

For example, HF STR-2000 shown at point A is the use of baseband analog to the commercial product of standard Marine AB for analog to digital conversion (ADC) with DSP in the TMS320C30 for high programmability. Commercial off the Shelf (COTS) cellular telephone handsets fall near B. Application specific Integrated Circuits provide processing capability. This is done through shifting of these designs towards the less programmable end of the axis. Digital cell set designs (C), in a similar fashion, rely heavily on digital filter ASICs for frequency translation and filtering, despite their accessing the spectrum at IF. SPEAK easy II, provides a GFLOP of programmable DSP. This it does by shifting this implementation to the right. The virtual radio (V) delivers a single channel radio using a general purpose processor, DEC's (Compaq's) Alpha. Point (X) is considered as the ideal software radio with digital RF, with all functions programmed on a general purpose processor. Currently, such designs are highly impractical economically even considering the presence of maximum flexibility in them and being the focus for a big research activity.

This parameter space differentiates software radios (V)–(X) quantitatively from programmable digital radios (PDR's) (A)-(O) which may have more than one RF band mode. Achievement of programmability has become a reality. This is via baseband DSPs. PDR is not a software radio without digital IF. PDR hardware modules, "slices" have to be interchanged for changing RF bands. Any such slice radio is a hardware defined one but not an SDR. On the other hand, a multi-slice radio with all slices in it and the capability of getting selected by software is an SDR. In an SDR transmitter, baseband signals are transformed into sampled waveforms via channel modem functions implemented in the software that drives high performance DAC's. Pre-emphasis or non-linearly pre-coding may be done for these signals using the IF processing software.

A PDR is not an ideal software radio if any crucial aspect of the channel waveform is implemented using programmable hardware (like e.g. voltage controlled oscillator), rather than using a software (like e.g. sine/cosine look up table). SDR's of the current generation are evolving in the direction of the ideal software radio with advancement in technology. SDR architecture should accept this evolution, and technology tradeoffs mentioned below.

2.8 Software Communications Architecture (SCA)

The Software Communications Architecture (SCA), used in conjunction with Software Defined Radio SDR, facilitates provision of portability of certain elements and provides a common interface for easy compilation of different modules written using eclectic sources. The SCA emerged as the result of the implementation of the Joint Tactical Radio System (JTRS) technique. Since this technique called for the compilation of resources from various suppliers and also the reuse of software was to be considered in JTRS, the SCA was defined and implemented accordingly.

By enabling the SDR to load waveforms, run applications and getting networked into a system, the SCA aims at controlling the structure and operation of the SDR. In a high degree of interoperability is ensured by SCA since software used to generate a particular waveform can be imported into several other radio systems. This facilitates the offsetting differences present in the different sources.

2.8.1 SCA Basics

The definition of the components of SDR, and, in particular, the description of the interfaces, is carried out by the SCA. The following are the two main advantages of SCA:

- The software elements or modules can be written in different organizations and can be compiled as one.

- The interoperability of the modules can be significantly improved facilitating increased cost savings since SCA enables the re-use of modules.

The failure of SCA can be identified in three main headings and there is a need to categorize these failures to enable planning an optimal solution for these problems of failure. The categories are:

- ***Management***: Managing the software is done by the SCA which includes various applications like plug and play, deployment and configuration software.
- ***Node***: Bootstrapping and access to hardware are some of the applications that come under the SDR.
- ***Application***: Waveform generation, demodulation, frequency translations are some of the signal processing activities carried out by the software.

The application category alone is reusable in different platforms while the other two are platform specific. Yet this portability has the advantage of significant cost savings. Designing specific software requires many years of training, capability to make the software work in spite of inconsistencies in the different platforms and rigorous training to ensure good design. This is highly expensive.

2.8.2 COBRA

COBRA – Common Object Request Broker Architecture is a middleware, which is an integral part of SCA software that enables inter-module communication i.e., the different modules and software elements can be compiled together. Object Management Group (OMG) defines COBRA with an aim to ensure interoperability between varieties of software modules.

2.8.3 SCA Compliance and Testing

SCA compliance is an important factor to be considered for SCA software, which determines the compliance of Application Programming Interface to other SCA compliant software. In addition to compliancy, this is also tested for perfect operability. Large SDR projects can ideally employ SCA since it provides a strong interface where communication is reliably carried out in a known standard format. SCA is not recommended for smaller projects since there is an overhead associated with it. Whatever the decision may be, i.e.

whether to implement SCA or not should be taken at the beginning of the design phase itself.

Chapter 3 provides a description of the basic characteristics of cognitive radioalong with key aspects like: Cognitive cycle, varieties of Cognitive radios and Cognitive radio network and its importance in next generation wireless architecture are detailed.

3

Cognitive Radio

3.1 Introduction

Wireless communication and the utilization of the radio frequency spectrum have witnessed a tremendous increase during the past few decades. The multitude of different wireless devices and technologies, the dramatic boom in the number of wireless subscribers, the continuous demand for higher data rates and the advent of new applications are all reasons for the radio frequency spectrum becoming more and more crowded.

Efficiency and reliability can be provided through the use of opportunistic radio resource utilization based on cognitive radio (CR) which also improves spectrum efficiency. Most of the IoT technologies like RFID, IEEE 802.15.4 (ZigBee) use the UHF and ISM frequency bands that are always getting overcrowded. The same spectrum being shared with other networks like WLAN may result in interference and congestion. Efficient use of the spectrum through intelligent and dynamic management is possible through use of CR. But the employment of CR in IoT requires exploitation of more research directions for handling heterogeneity, providing energy efficiency and using cross layer strategies. In the matter of self organization in IoT, it is possible for CR to enable nodes to manage communication by themselves. It is also possible for CR to obviate collision and wastage of valuable resource. Detection of the radio environment is done by CR through finding a solution for reuse of frequency space time avoiding collision. Such capabilities have propped up CR as a promising enabler for IoT and M2M.

3.2 Understanding of Cognitive Radio

The ultimate objective of CR is to address the spectrum inefficiency problem. The radio frequency spectrum is divided into frequency bands that are then allocated to different systems. The allocations are decided by the regulatory authorities in each country such as the Federal Communications Commission (FCC) in the United States. Most of the spectrum has already been assigned to different systems. Such allocations vary from country to country. Any wireless system ushered in needs the availability of a frequency band with the requirement of worldwide collaboration. Hence the current frequency allocation is rigid and inflexible despite its assurance of low interference considering the operation of each system in a different band. This has resulted in an apparent spectrum scarcity that realizes as heavy congestion in certain frequency bands. However, many of the frequency bands have been allocated to legacy systems that are rarely used or to systems whose degree of frequency band utilization varies sharply from time to time and location to location. Hence, a significant amount of spectrum is still available. But there is no possibility of its exploitation fast on the basis of the fixed frequency allocation policy. The consequence is inefficient use of the radio frequency spectrum depending on time, frequency band and location. Thus, spectrum sensing is becoming an increasingly important functionality to modern and future wireless communication for identifying underutilized spectrum and characterizing interference, and consequently, achieving reliable and efficient operation.

A cognitive radio is characterized as a radio that is aware of its environment, and the internal state and then plan, decide and act on those conditions. The system can learn from these adjustments and use them to for making future choice, all while considering the end-to-end objectives. In general, the cognitive radio involves ability to identify free spectrum, channel occupancy, type of data transmitted through the channel, modulation type used etc., Apart from that, it should have knowledge of the regulatory requirements and knowledge of geography of that location.

In a few occurrences it might be important to utilize a software defined radio, with the goal of that it can reconfigurable itself to meet the ideal transmission innovation for a given arrangement of parameters. As needs be Cognitive radio innovation and software defined radio are regularly firmly connected.

3.3 Cognitive Radio Architecture

The important design part in cognitive radio architecture is RF areas should be adaptive in nature. It has the ability to swap the frequency bands and change in transmission modes that occupy different bandwidths. To accomplish the required level of performance will require an exceptionally adaptable front end. Conventional front end technology can't deal with these necessities, because they are bandlimited. Appropriately, the required level of execution must be accomplished by changing over to and from the sign as close to the radio antenna as could be expected under the circumstances. No simple signal processing is required along these lines, all the processing being taken care of by the digital signal processing.

The transformation to and from the digital arrangement is taken care of by digital-to-analog converters (DACs) and analog-to-digital converters (ADCs). The accomplishment of the execution required for a cognitive radio requires DACs and ADCs to have a dynamic range and also the capacity to work over a wide range, reaching out to numerous GHz. However they should have the capacity to handle significant levels of power considering the transmitter.

The advancements in VSL technology have enabled the availability of the required ADC and DAC technology in the near future, thereby making cognitive radio a reality.

3.4 Cognitive Radio Characteristics

Cognitive radio characteristics are defined by various stages of cognitive cycle. Figure 3.1 shows the cognitive cycle. The first step in cognitive cycle is to '*OBSERVE*' the outside world and '*LEARN*' the present condition. The second step is to identify the opportunities (spectrum holes). The Third phase

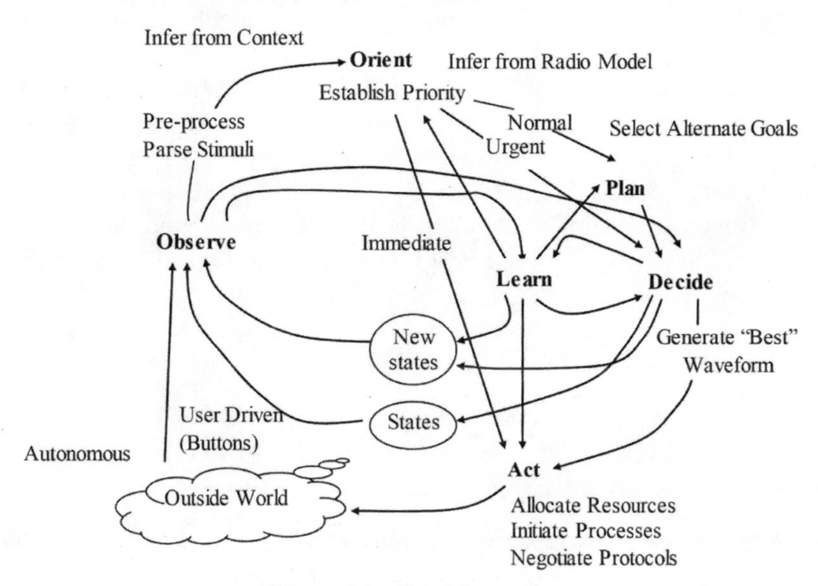

Figure 3.1 Cognitive cycle.

is to '*PLAN*'. The identified spectrum holes are to be allocated to the CR users. This accomplishes the *PLAN* and *DECIDE* step. The next step is to '*ACT*'. The allotted secondary CR user transmits its signals using the allotted spectrum. This changes the situation in the outside world, the process gets restarted with the first step and the cycle continues. Sometimes, it can also allocate the unused spectrum to the secondary users by prioritizing the secondary users. This stage is *ORIENT* where the secondary users are prioritized before allocating the spectrum. Cognitive radio characteristics are listed below

- Intelligent radio that uses spectrum licensed to other users when they are not using it. This is known as 'Bandwidth Harvesting'.
- It is a **software-designed radio** with cognitive software.
- CR can **sense** the environment.
- CR **adapts** its way of communication to minimize the caused interference.

3.4.1 Primary and Secondary Users

One of the most important components of cognitive radio concept is its ability to measure, sense, learn, and be aware of the parameters related to the radio channel characteristics, availability of spectrum and power, interference and noise temperature, operating environment of the radio, user requirements and applications, available networks (infrastructures) and nodes, local policies and other operating restrictions. In cognitive radio terminology, *primary users* (PU) can be defined as the users who have higher priority or legacy rights on the usage of a specific part of the spectrum.

Secondary users (SU) have lower priority and exploit this spectrum in such a way that they do not cause interference to primary users. Therefore, secondary users require CR capabilities, such as sensing the spectrum reliably to check whether it is being used by a PU and to change the radio parameters to exploit the unused part of the spectrum. The 3D view of the CR environment in time, power and frequency axes is shown in the Figure 3.2. The gray blocks display the frequencies used by the primary users. The white blocks in between displays the unused frequencies. These white blocks are nothing but spectrum holes.

3.5 Cognitive Radio Environment

The various steps involved in the creation of a cognitive radio environment are listed below:

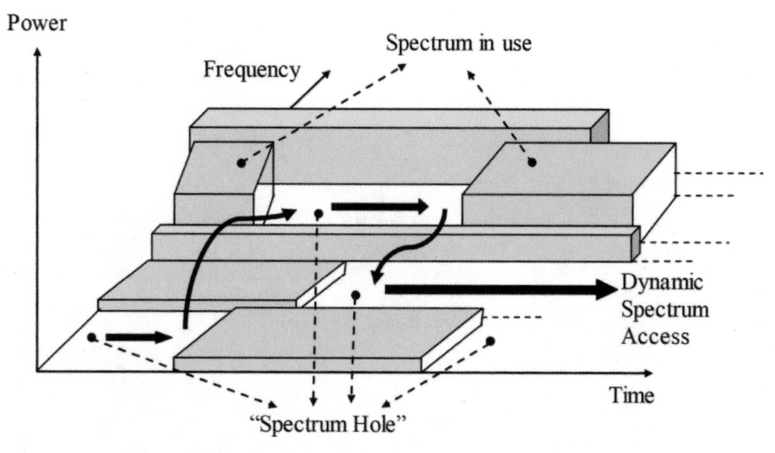

Figure 3.2 Overview of spectrum in CR environment.

Step 1:

To start with, there are two PUs using the spectrum. This is indicated by the increase in amplitude of the signal in the graph. In Figure 3.3, the PUs occupies 1 MHz and 5 MHz frequencies. This indicates the absence of use of frequencies 2 MHz, 3 MHz and 4 MHz and the ability to get allocated to the SU's.

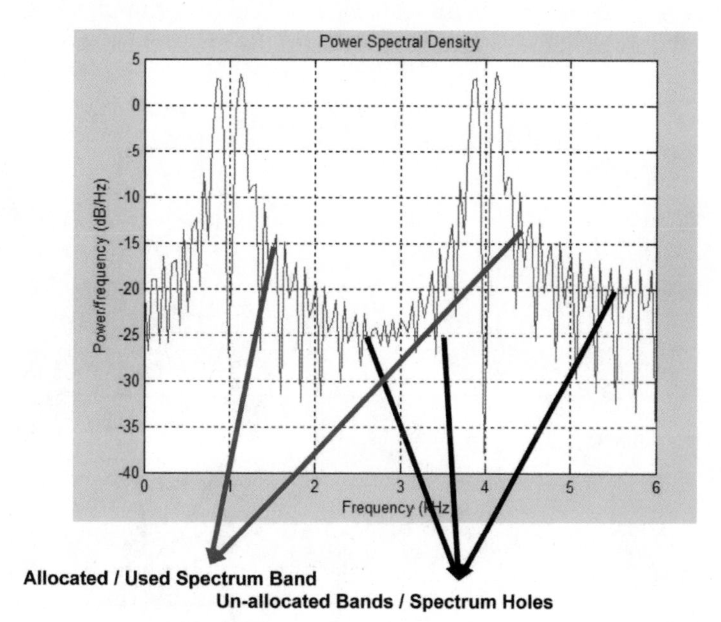

Figure 3.3 Primary user occupancy.

In this case, the presence of the primary user is indicated by taking the amplitude modulation of the signal.

Step 2:
One of the SUs wants to use the spectrum if free. Since 2 MHz, 3 MHz and 5 MHz frequency are free, one of the frequencies can be allocated to the secondary users can be allocated. Figure 3.4 shows the allocation of the secondary user for using 2 MHz frequency. This enables the effective use of the unused spectrum.

Step 3:
In this step, the PU returns to use the frequency 2 MHz which has been allocated to the secondary user in the previous step. Hence the SU should be suppressed to enable the use of the frequency by the PU. As shown in Figure 3.5, the SU gets suppressed when the PU returns to use the spectrum.

3.6 Types of Cognitive Radios

The various types of cognitive radios are policy CRs, procedural CRs and ontological CRs. Policy CRs are governed by the radio policy which has a set of rules based on the location and the environment of the radio, constraints from primary users of the spectrum etc., There is need for implementation of a rule based domain knowledge. This is due to the requirements of spectrum

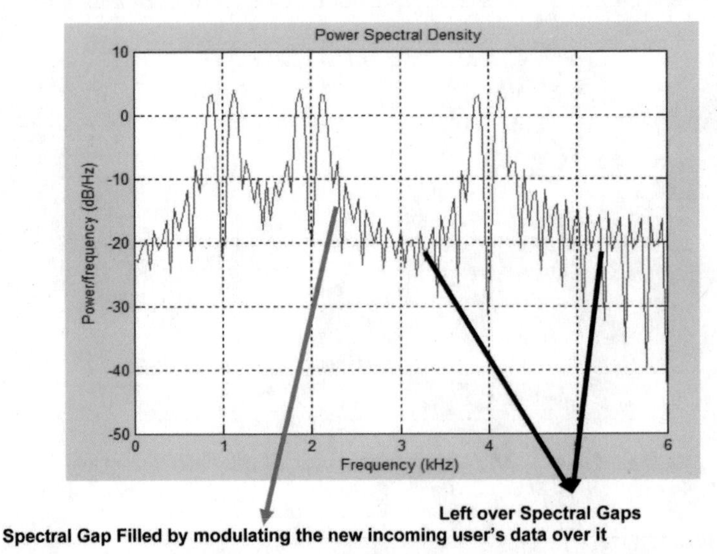

Figure 3.4 Presence of secondary user.

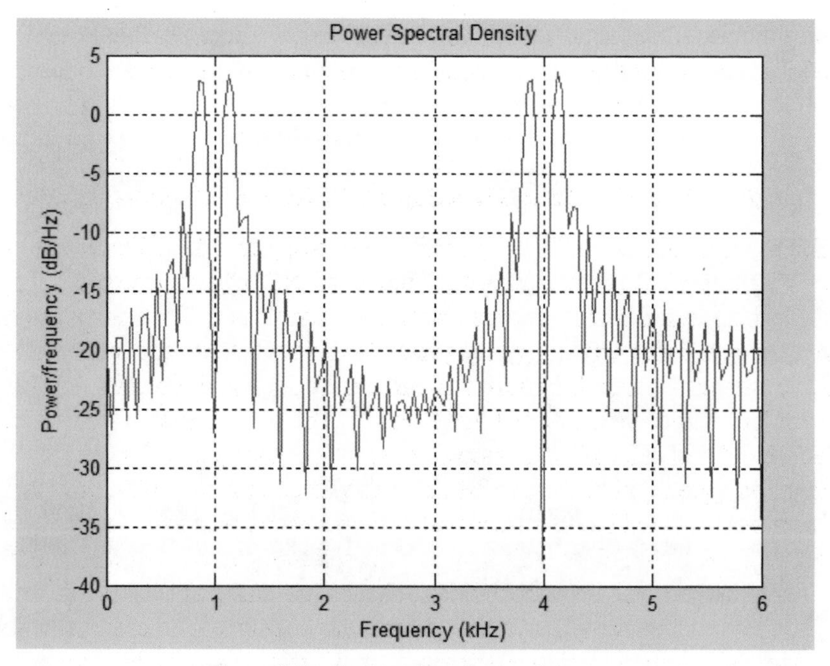

Figure 3.5 Suppression of secondary user.

regulations of licensed primary users not getting affected by unlicensed CRs. Policy holders usually do not have a learning or reasoning engine.

3.6.1 Procedural CRs

They rely on observations through the use of hard-coded algorithms that lay down the various actions needed for different inputs to enable adaptation to operational parameters. Adaptive actions arise from observations based on procedural knowledge which has its description done by a set of pre-defined hard coded function. As a result, procedural radios have higher flexibility compared to the policy radios. On the other hand, they are less intelligent as they operate in a deterministic manner. Predictable actions are their inputs with specific combinations of observations made as for example, the use of a hard coded genetic algorithm which provides use to adaptive actions from the observations of the algorithms may be made.

3.6.2 Ontological CRs

These are considered to be the most flexible and intelligent radios for the reason they use both the reasoning and learning engines Radio Knowledge

Representation Language (RKRL) is among the solutions which ontological reasoning uses for describing the existence of entities and relationships inter se, and the method used for subdivision among them on the basis of similarities and differences. Ontologies help the reasoning engine in the inference of radio frequency environment and make intelligent decisions. Ontological CRs are known to be proactive. This is because they use their own reasoning based on updated knowledge base (Cognitive cycle) instead of a predetermined logic for inferring the subsequent action.

3.7 Cognitive Radio Networks

Cognitive radios form a cognitive radio network (CRN), by extending radio features to the network layer and above. CRNs are networks known for their heterogeneity. They comprise networks and communication systems of different kinds. As a general rule, CRNs have the ability of deployment in network centric, distributed, ad-hoc and mesh architectures. The two categories of CRNs based on the presence of an infrastructure support are as follows:

3.7.1 Infrastructure (Centralized) CRNs

In this type, there exists a central authority for purpose of monitoring the administration of an infrastructure CRN, as for example, IEEE 802.22 wireless regional area network (WRAN) which is an infrastructure based CRN similar to a cellular network which comprises a base station (BS) and consumer premise equipments (CPEs). The BS acts as a data fusion centre for the spectrum sensed data reported by CPEs. Allocation of uplink and downlink channels to CPEs is done by BS in its cell on the basis of the information gathered. There is another example of such a network which is an access point with a set of cognitive radio enabled nodes associated with it like an IEEE 802.11 network, but in a situation where nodes are unlicensed.

3.7.2 Ad-hoc Mode CRNs

The working of CRN without infrastructure has no dedicated fusion centre or channel allocation authority. The function of this network is similar to that of an ad-hoc network. Here, cognitive radios make independent decisions on the subject of channel access due to the absence of a central authority. Other subjects on which such independent decisions are made include transmission power and routing. CRNs can also be deployed in mesh architecture by combining infrastructure based and ad hoc modes with the help of wireless connections between BS, in the same way as the Hybrid Wireless Mesh Networks.

3.8 How Cognitive Radio Empowers Internet of Things

The potential benefits brought to IoT through application of various mechanisms and techniques for CR discussed in the earlier sections for showing how CR empowers IoT are summarized in this section. The parameters included automaticity, scalability, tackling heterogeneity, energy efficiency, etc. this was done for providing a global picture of various CR approaches that can be applied in the backdrop of IoT. Considering CR-IoT concepts are relatively new, only very few topics are seen in the works related to CR-IoT. In addition, a few discussions of a general nature a self organization, energy efficiency etc., by the addition indicates their relevance to IoT. The objective was to demonstrate IoT getting intelligent; empowering of IoT by CR. Spectrum becoming scarce is common knowledge. Trillions of devices are expected to get wireless connection in the near future. Quality of Service, reliability, scalability and energy efficiency are issues that require attention. The author's consideration is that application of CR in the context of IoT will empower IoT. This is because CR allows efficient spectrum utilization, higher accessibility, greater ease in use, better adaptability, improved inter connectivity, increased scalability and empowered reliability. CR brings in several benefits to IoT the first of which is it enables effective spectrum utilization. This aspect has been covered in detail in the previous sections.

Other significant benefits are detailed below:

- Greater accessibility: The user uses a single device for accessing various networks and services. So, he specifies his requirements. CR scans for the services available and indicates the options available to the user. Trillions of devices available in IoT are expected to get wireless communication. Application of CR to IoT will improve accessibility to various networks and services.
- Better usability: Objectives and preferences of the user are within the knowledge of any CR enabled device. Further, the device is capable of adaptation of its operation to simplification of tasks. For instance, any CR enabled device can switch over from cellular network to WiFi network known for lower cost and higher user performance, soon after the availability of the latter, thereby enabling greater ease for the user.

- Improved adaptability: Any CR enabled device has the ability of automatic adaptation to local environment. This is of inestimable use when the user roams across borders, considering the ability of the device to adapt itself for compliance with local radio operations.
- Better interconnectivity: A large number of devices are found in IoT. These include laptops, smart phones, game consoles, media players etc., which are expected to get wireless connection. CR enables efficient communications among multi terminal/multi frequency devices improving interconnectivity among trillions of devices which expect wireless connection in IoT.
- Scalability: CR enabled devices are known for communication inter se in the form of collaboration among neighbour devices and, as a result, the network can scale to a much larger number of users.
- Enhanced reliability: It is possible for self figuring mesh wireless network to do self figuring with cognitive radio. This will help them to obviate disruption or failure through re-routing around node failures or congestion areas. More robust and reliable communications is therefore possible through CR.

3.9 Challenges

3.9.1 Spread Spectrum Primary Users

Primary users who use frequency hopping and spread spectrum signaling, wherein the power of the PU signal is distributed over a wider frequency, despite the actual information bandwidth being much narrower, are difficult to detect. Especially, frequency hopping based signaling creates significant problems in the matter of spectrum sensing. This problem can be avoided to some extent if the hopping pattern is known and perfect synchronization to the signal can be achieved.

3.9.2 Hidden Node/Sharing Issues

Hidden PU problem is similar to the hidden node problem in the carrier sense multiple accessing (CSMA). This problem can be caused by many factors including severe multipath fading or shadowing that secondary users observe while scanning primary users' transmissions. Figure 3.6 in an illustration of a hidden node problem. Here, CR causes unwanted interference to the PU (receiver) due to the inability to detect primary transmitter's signal arising of the positioning of the devices in space.

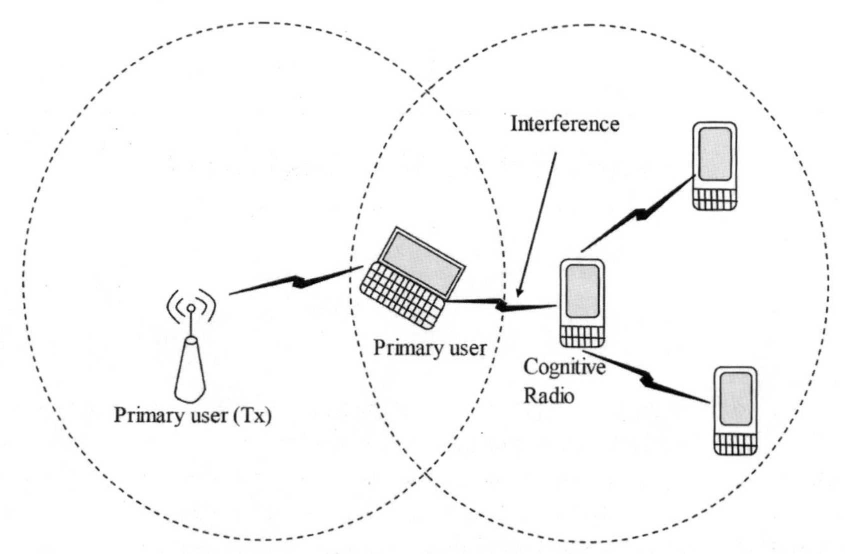

Figure 3.6 Illustration of hidden primary user problem.

3.9.3 Sensing Time

A primary user can claim its frequency bands anytime while cognitive radio is operating at that band. In order to prevent interference to and from primary license owners, cognitive radio should have the ability to identify the presence of primary users as quickly as possible and should vacate the band immediately. Hence, the sensing method should identify the presence of a primary user within a specific duration. This requirement has a limit on the performance of sensing algorithm and creates a challenge for cognitive radio design.

3.9.4 Other Challenges

Some other challenges that need to be considered while designing effective spectrum sensing algorithm include implementation complexity, presence of multiple secondary users, coherence times, multipath and shadowing, cooperation, competition, robustness, heterogeneous propagation losses, and power consumption.

To sum up, Cognitive radio technology provides four specific facilities for users. They are, Spectrum Sensing, spectrum management, spectrum mobility and spectrum sharing. In the chapters that follow, the authors present a detailed discussion of the above topics, status from spectrum sensing.

4

Next Generation Networks

4.1 Introduction

Next generation (xG) communication networks go by the names of dynamic spectrum access Networks (DSANs) and cognitive radio networks. They provide the facility of high bandwidth to users via, heterogeneous wireless architectures and dynamic spectrum access techniques.

Cognitive radio is the significant enabling technology of xG networks. The use or sharing of spectrum in an opportunistic manner is possible thanks to cognitive radio techniques. Operation of cognitive radio is possible in the best available channel with the help of the dynamic spectrum access. Cognitive radio technology provides four specific facilities for users. These are

1. Spectrum Sensing – Involves detection of unused spectrum, detection of the presence of licensed users during the operation in a band, without any harm to other users.
2. Spectrum management – Selection of the best available channel for meeting the communication needs for a user.
3. Spectrum mobility – Involves vacation of the channel by the SU on the return of the licensed user, thereby answering communication requirements during the transition from one spectrum to another better one.
4. Spectrum sharing – Coordination of access to a channel with other users. This provides the facility of a fair spectrum sharing among coexisting xG users.

Details of xG network components and intersections between them are provided in Figure 4.1. Enhancement of spectrum efficiency is seen as cooperation between spectrum management and spectrum mobility functions with application, routing, transport, medium access, and physical layer being carried out.

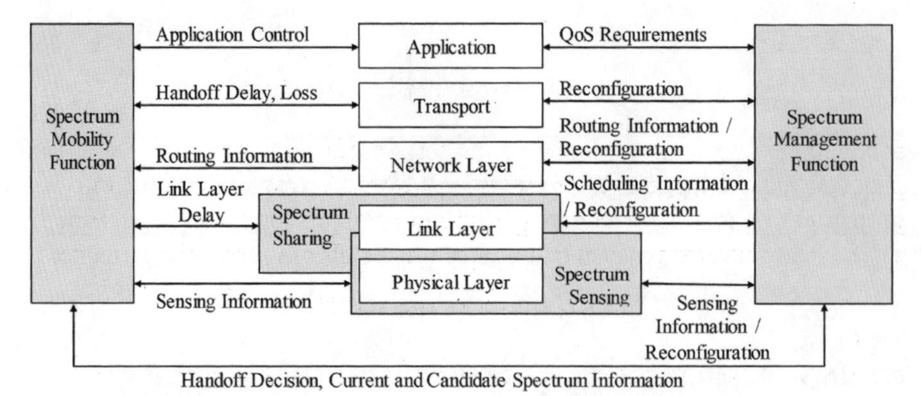

Figure 4.1 Next generation network communication functionalities.

4.2 Classical Hypothetical Analysis of Spectrum Sensing

Spectrum sensing is the most important key element in Cognitive Radio. It enables the CR to adapt to its environment by detecting spectrum holes. In CR, the PU has higher priority or legacy rights on the usage of the spectrum. Figure 4.2 shows the principle of spectrum sensing.

In the Figure 4.2, the PU transmitter transmits data to the PU receiver in a licensed spectrum band, while a pair of SUs tries to access the spectrum. Performance of spectrum sensing by the SU transmitter is needed for protecting PU transmission and to detect and as certain the presence of a

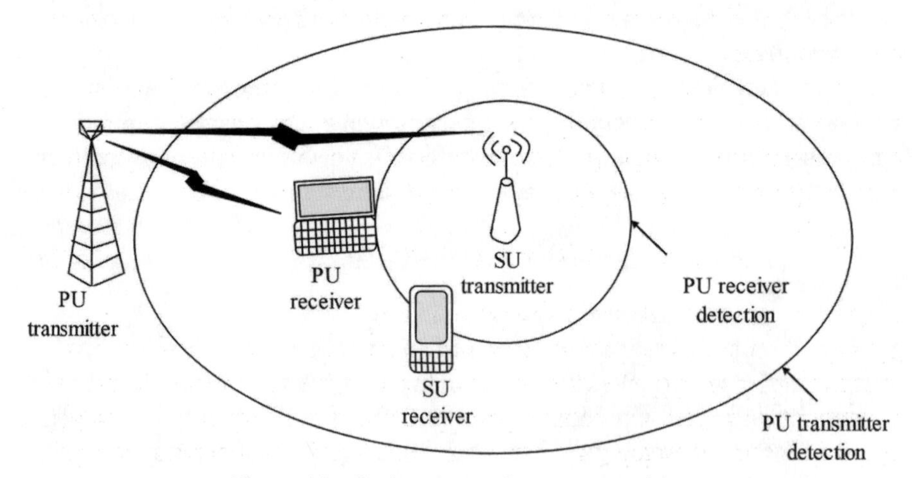

Figure 4.2 Basic concept of spectrum sensing.

PU receiver in the coverage of the SU transmitter. The spectrum sensing scheme basically employs transmitter detection, involving the determination of the frequency at which the transmitter is operating. A hypothesis model for transmitter detection is described later. In general, it is difficult for the SUs to differentiate the PU signals from other pre-existing SU transmitter signals. Therefore, all are treated as one received signal, $s(t)$. The received signal at the SU, $x(t)$ can be expressed as $x(t) = n(t)H_0$

$$x(t) = s(t) + n(t)H_1 \tag{4.1}$$

where $n(t)$ represents AWGN. H_0 and H_1 represent the hypothesis of the absence and presence of PU signals respectively. The performance of detection algorithm depends on some important parameters: the probability of detection (Pd), the probability of false alarm (Pf), and the probability of missed detection (Pm).

The total error rate is the sum of the probability of false alarm Pf and the probability of missed detection Pm or $(1 - Pd)$. Thus, the total error rate is given by:

$$Pe = Pf + Pm = Pf + (1 - Pd), \tag{4.2}$$

where $(1 - Pd)$ shows the probability of missed detection (Pm). The sensitivity of system depends on the probability of detection (Pd) while the specificity of the system depends on the probability of false alarm (Pf) and the probability of missed detection (Pm).

Pd is the probability of precise detection of the PU signal present in the considered frequency band. In terms of hypothesis, it is given as:

$$Pd = \Pr(signal\ is\ \det ected|H_1) \tag{4.3}$$

Pf is the probability that the detection algorithm falsely decides that PU is present in the scanned frequency band when it actually is absent, and it is written as

$$Pf = \Pr(signal\ is\ \det ected|H_0) \tag{4.4}$$

The goal is to maximize Pd and minimize Pf to improve system performance. Probability of missed detection Pm is the complement of Pd. Pm which indicates the likelihood of not detecting the primary transmission when PU is active in the band of interest and can be formulated as:

$$Pm = 1 - Pd = \Pr(signal\ is\ not\ \det ected|H_1) \tag{4.5}$$

Total probability of making a wrong decision on spectrum occupancy is given by the weighted sum of Pf and Pm. Hence, the key challenge in the transmitter detection approach is to keep both Pf and Pm under control considering high Pf corresponds to poor spectrum utilization by CR and high Pm may result in increased interference at primary user. There are two basic hypothesis testing criteria for testing a hypothesis in spectrum sensing: the Neyman-Pearson (NP) and Bayes tests. The NP test aims at maximizing Pd (or minimizing Pm) under the constraint, $Pf \leq \alpha$, where α is the maximum false alarm probability, while, the Bayes test that minimizes the average cost is given by Expression (4.6)

$$R = \sum_{i=0}^{1} \sum_{j=0}^{1} C_{ij} P_r(H_i/H_j) P_r(H_j) \tag{4.6}$$

Where C_{ij} are the costs of declaring H_i when H_j is true, $P_r(H_i)$ is the prior probability of hypothesis H_i and $P_r(H_i/H_j)$ is the probability of declaring H_i when H_j is true. Both of the tests are equivalent to the likelihood ratio test (LRT) given by:

$$\Lambda(x) = \frac{P(x/H_1)}{P(x/H_0)} \tag{4.7}$$

where $P(x(1), x(2), \cdots, x(M)|H_i)$ is the distribution of observations $x = [x(1), x(2), \cdots, x(M)]^T$ under hypothesis H_i, i belongs to $\{0, 1\}$, $\Lambda(x)$ is the likelihood ratio. The distributions of $P(x|H_i)$ in both the tests are known. When there are unknown parameters in the probability density functions (PDFs), the test is called composite hypothesis testing. Generalized likelihood ratio test (GLRT) is one kind of the composite hypothesis test. In the GLRT, the unknown parameters are determined by the maximum likelihood estimates (MLE). The choice between the two hypotheses is made by comparing a test statistic T with a threshold γ.

The probability of false alarm and detection are given by the equations

$$Pf = P(T > \gamma|H_0) \& Pd = P(T > \gamma|H_1) \tag{4.8}$$

Spectrum sensing techniques can be categorized as transmitter detection, cooperative detection, and interference based detection. This is shown in Figure 4.3. The sections that follow provide a description of each of these sensing methods for xG networks and a description of open research topics in this area.

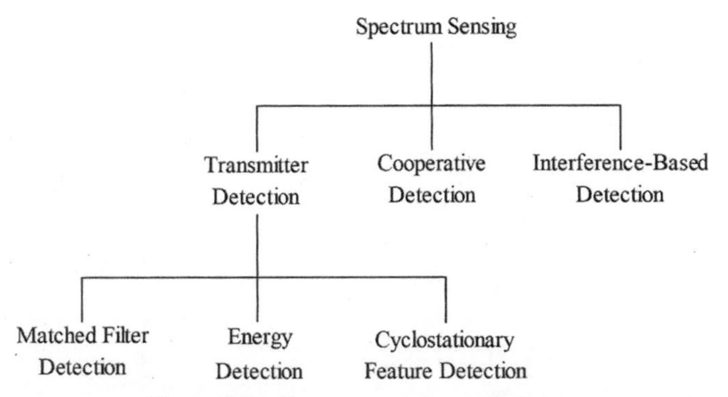

Figure 4.3 Spectrum sensing techniques.

4.3 Transmitter Detection (Non-Cooperative Detection)

Making a distinction between used and unused bands is an important requirement for the CR, which should, therefore be capable of determining the local presence of a signal in a certain spectrum. The basis of transmitter detection is the detection of the weak signal from a primary transmitter through the local observations of xG users.

The basic hypothesis model for transmitter detection can be defined as

$$x(t) = \begin{cases} n(t) & H_0, \\ hs(t) + n(t) & H_1 \end{cases} \tag{4.9}$$

Where $x(t)$ the signal is received by the xG user, $s(t)$ is the transmitted signal of the primary user, $n(t)$ is the additive white Gaussian model (AWGN) and is the amplitude gain of the channel. H_0 represents a null hypothesis which postulates none of the licence user signals in a specific spectrum band. As against this, H_1 is an alternative hypothesis, indicating the existence of some licensed user.

Three schemes find use in transmitter detection on the basis of the hypothesis model. Investigation of matched filter detection, energy detection and cyclostationary feature detection techniques proposed for transmitter detection in xG networks is done in the subsections that follow.

4.4 Matched Filter Detection

The matched filter is the linear optimal filter used for coherent signal detection for maximizing the signal-to-noise ratio (SNR) in the presence of additive stochastic noise. As shown in Figure 4.4, it is obtained by correlating a

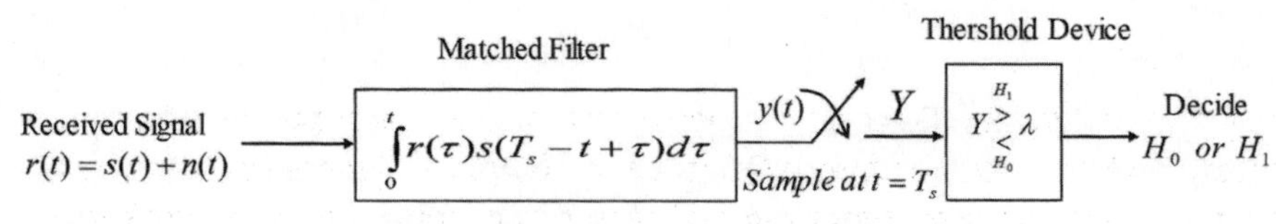

Figure 4.4 Matched filter detection.

known original PU signal $s(t)$ with a received signal $r(t)$ where T is the symbol duration of the PU signals. The output of the matched filter is then sampled at the synchronized timing. If the sampled value Y is greater than the threshold λ, the spectrum is determined to be occupied by the PU transmission. This detection method is known as the optimal detector in stationary Gaussian noise. It shows a fast sensing time, which requires $O(\frac{1}{SNR})$ samples to achieve a given target detection probability. The matched filter necessitates not only a priori knowledge of the characteristics of the PU signal but also the synchronization between the PU transmitter and the CR user. The performance of the matched filter is dismal when this information is not accurate. Furthermore, CR users need to have different multiple matched filters dedicated to each type of the PU signal, which increases the implementation cost and complexity.

4.5 Energy Detection

When the receiver is unable to procure adequate information on the primary user signal (as, for example, when the power of the random Gaussian noise is known only to the receiver) the optimal detector is energy detection. The measurement of the energy of the received signal requires the output signal of the bandpass filter with bandwidth W to be squared and investigation done over the observation interval T. At the end, the output of the integrator Y is compared with a threshold, λ for taking a decision on the presence or absence of a licensed user. Figure 4.5 illustrates energy detector.

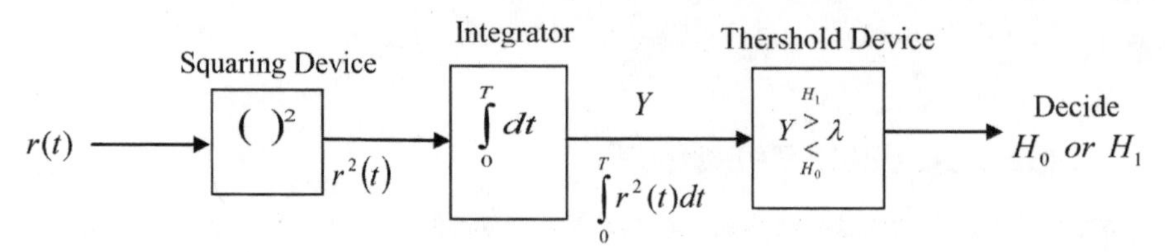

Figure 4.5 Energy detector.

In the event of application of the energy detection in a non-fading environment where h is the amplitude gain of the channel, the probability of detection P_d and false alarm P_f are given as,

$$P_d = P\{Y > \lambda|H_1\} = Q_m(\sqrt{2\gamma}, \sqrt{\gamma}),$$

$$P_f = P\{Y > \lambda|H_0\} = \frac{\Gamma(m, \lambda/2)}{\Gamma(m)} \tag{4.10}$$

Where λ is the threshold SNR, $u = TW$ is the time bandwidth product, $\Gamma(\bullet)$ and $\Gamma(\cdot,\cdot)$ are complete and incomplete gamma functions respectively and $Q_m(\bullet)$ is the generalized Marcum Q function. A low P_d would mean missing the presence of the primary user with a high probability which increases the interference to the primary user. A high P_f would result in low spectrum utilization since false alarms are prone to increase missed opportunities.

When we consider the shadowing and the multipath finding factors are considered for the energy detector, while P_f is independent of Γ, when a variation in amplitude gain of the channel, h arises due to shadowing or fading, P_d provides the probability of detection conditioned on instantaneous SNR as follows:

$$P_d = \int_x Q_m(\sqrt{2\gamma}, \sqrt{\lambda}) f_\gamma(x) dx \tag{4.11}$$

Where $f_\gamma(x)$ is the probability distribution function of SNR under the fading process.

As against this, there is the fact that the performance of the energy detector is exposed to uncertainty in noise power. Solution to this problem requires the use of a pilot tone from the primary transmitter to assist improvement of the accuracy of the energy detector. There is also the inability of the energy detector to differentiate signal types but can locate the presence of the signal. The energy detector faces the problem, of proneness to false detection initiated by unintended signals.

4.6 Cyclostationary Feature Detection

Feature detection determines the presence of PU signals by extracting their specific features such as pilot signals, cyclic prefixes, symbol rate, spreading codes, or modulation types from its local observation. These features introduce

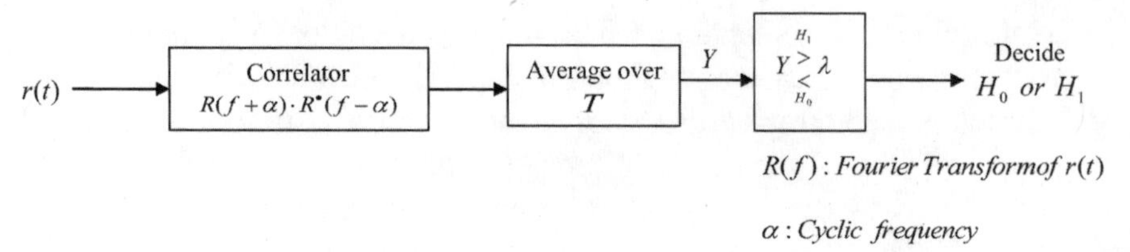

Figure 4.6 Cyclostationary feature detection.

built-in periodicity in the modulated signals, which can be detected through analysis of a spectral correlation function. The feature detection leveraging this periodicity is also called cyclostationary detection. Here, the spectrum correlation of the received signal $r(t)$ is averaged over the interval T, and compared with the test statistic for determining the presence of PU signals, similar to energy detection.

The main advantage of the feature detection is its robustness to the uncertainty in noise power. It can also distinguish the signals from different networks. This method allows the CR user to perform sensing operations independently of those of its neighbors without synchronization. It is shown in Figure 4.6.

4.7 Cooperative Detection

Under fading or shadowing, the received signal strength can be very low and this can prevent a node from sensing the signal of interest. Noise can also be a challenge when energy detection is used for spectrum sensing, despite the presence of spectrum sensing techniques that are robust in the presence of noise, such as cyclostationary feature detection. Detection of the signal of interest may not come through due to a low signal-to-noise ratio. The idea of cooperative spectrum sensing in a RF sensor network is the collaboration of nodes on deciding the spectrum band used by the transmitters emitting the signal of interest. Nodes send either their test statistics or local decisions about the presence of the signal of interest to a decision maker, which can be another node. The unwanted effects of fading, shadowing and noise can be minimized through this cooperation. This is because a signal that is not detected by one node may be detected by another. As the number of collaborating nodes increases, the probability of missed detection for all nodes decreases.

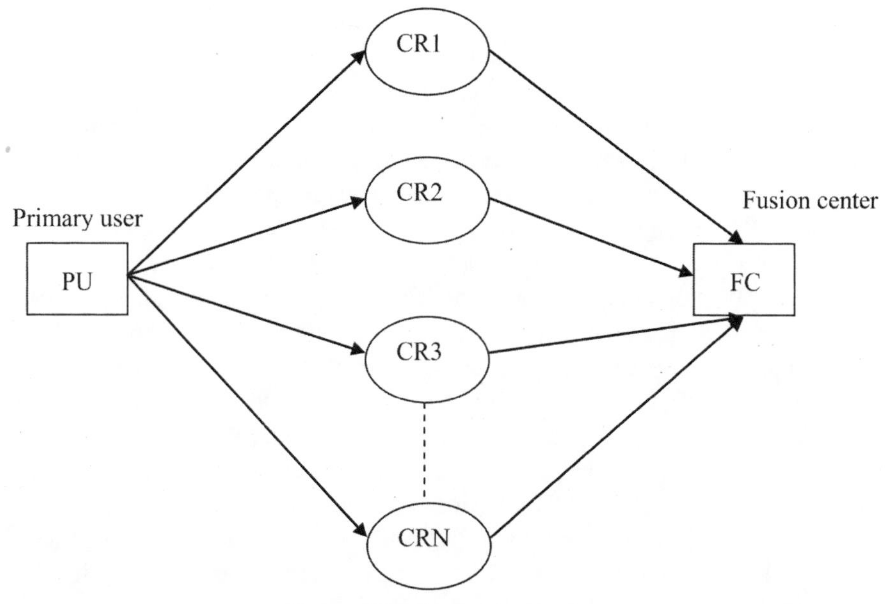

Figure 4.7 Cooperative sensing.

Cooperation in spectrum sensing also improves the overall detection sensitivity of a RF sensor network without the requirement for individual nodes to have high detection sensitivity. A smaller number of sensitive detectors on nodes mean reduced hardware and complexity. The trade-off for cooperation is more communication overhead. Since the local sensing results of nodes should be collected at a decision maker, where the decision is made, a control channel is required between the decision maker and the other nodes. There are three forms of cooperation in spectrum sensing: hard decision (also known as decision fusion), soft decision also known as data fusion) and quantized decision. The difference between these forms is the type of information sent to the decision maker. Three rules used by the decision maker under hard decision are logical–OR rule, logical-AND rule and majority rule. Cooperative sensing model is illustrated in Figure 4.7.

4.8 Interference Based Detection

A transmitter centric way regulates interference, implying the ability to control interference at the transmitter through radiated power together with out-of band emissions and location of individual transmitters. But, in practice, interference occurs at the receivers. A new model for measuring interference has been introduced. This is called interference temperature. The model points to the signal of a radio station designed for operation in a range indicating the approach of recent power to the level of the noise floor. With the appearance of additional interfering signals, there is increase in the noise floor at various points within the service area. This is indicated by the peeks above the original noise floor. As against this the traditional transmitter centric approach is able to manage interference at the receiver using the interference temperature limit, represented by the volume of new interference that could stand toleration from the receiver. The interference temperature model has the responsibility for cumulative RF energy from multiple transmissions, setting a cap on their aggregate level. xG users can use the spectrum band to the extent that they do not exceed this limit by their transmission some limitations on measuring the interference temperature that still exists. Interference refers to the expected fraction of primary users with disruption of service caused by xG operations. Factors inclusive of the type of unlicensed signal modulation, antennas, ability to detect active licensed channels, power control and activity levels of licensed and unlicensed users are considered by this method. Description of the interference disrupted by a single xG user is provided by this model which does not consider the effect of multiple xG users. There is also the total absence of knowledge on the location of nearby users on the part of xG users causing inability measurement of the actual interference through use of this method.

4.9 Neyman Pearson Fusion Rule for Spectrum Sensing in Cognitive Radio

A high dependence of the general performance on the fusion scheme is seen. It is, therefore, important to know in what manner is the local data from the SUs for improving the local sensing performance to the extent possible. Since in a CR network, a larger probability of detection (P_D) leads to less interference with PUs, and smaller probability of false alarm (P_{FA}) results in higher spectrum efficiency, it is desirable to maximize the P_D while P_{FA} is minimized. Nonetheless, the inability to carry out this optimization and P_D and P_{FA} simultaneously is shown as also the fact of the Neymann Pearson (NP) criterion being a good candidate for maximizing P_D forcing a restriction on P_{FA}.

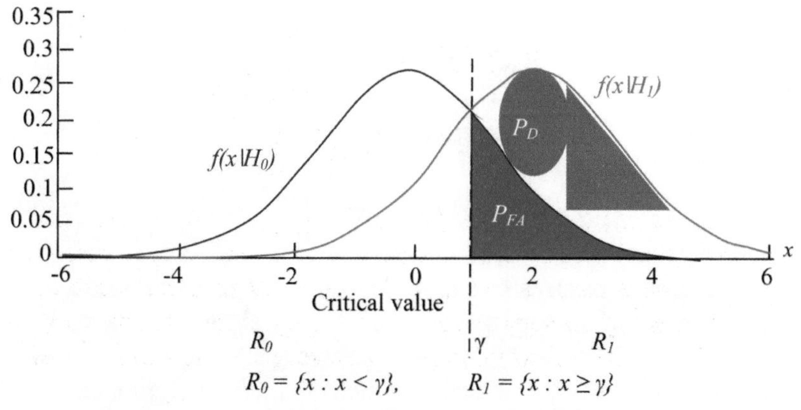

Figure 4.8 Observation space.

From the Figure 4.8, probability of errors and correct decisions are defined as

$$\text{Pr}obability\ of\ Flase\ Alarm = P_{FA} = \int_{\gamma}^{\infty} f(x|H_0)dx$$

$$\text{Pr}obability\ of\ Dismissal = 1 - P_{FA} = \int_{-\infty}^{\gamma} f(x|H_0)dx$$

$$\text{Pr}obability\ of\ Detection = P_D = \int_{\gamma}^{\infty} f(x|H_1)dx$$

$$\text{Pr}obability\ of\ Flase\ Dismissal = 1 - P_D = \int_{-\infty}^{\gamma} f(x|H_1)dx \quad (4.12)$$

System model

A parallel distributed detection system consisting of N secondary detectors with a fusion center in a CR network is considered. The spectrum sensing problem at each detector can be formulated by the following binary hypothesis test,

$$\begin{cases} H_0 : y_i = n_i, & i = 1, 2, , N \\ H_1 : y_i = s_i + n_i, & i = 1, 2, , N \end{cases} \quad (4.13)$$

where, s_i is the PU signal received at the ith SU. We also assume that the PU signals s_i and noise received at each SU are independent.

Hypothesis test at the fusion center

The final decision is to be made at the fusion center on the basis of the information, received Spectrum sensing can, therefore, be stated as a binary hypothesis testing problem, with the null and alternative hypotheses

$$\begin{aligned} &H0 : PU\ signal\ is\ absent \\ &H1 : PU\ signal\ is\ present \end{aligned} \tag{4.14}$$

The optimal decision rule at the fusion center in the sense of maximizing P_D for a given P_{FA} is the NP test. The structure of the fusion rule is obtained independently from the values of the local decisions u_i however the final decision is made upon the local decisions.

NP test is done at the fusion center for making the final decision. The NP formulation can be stated as follows: for a prescribed bound on the probability of false alarm at the fusion center, P_{FA} find the optimum decision rule which maximizes the probability of detection P_D,

$$\begin{aligned} P_d(\overline{d}_{NP}) &= \sum_{L(u)>\eta} P(u|H_1) + \gamma \sum_{L(u)=\eta} P(u|H_1), \\ P_{fa}(\overline{d}_{NP}) &= \sum_{L(u)>\eta} P(u|H_0) + \gamma \sum_{L(u)=\eta} P(u|H_0), \end{aligned} \tag{4.15}$$

where $u = [u_1, \cdots, u_N]^T, \overline{d}_{NP}(u)$ is the conditional probability of accepting H_0, $\gamma \leq 1$ is the randomization constant and the decision threshold, η are determined on the basis of the desired false alarm probability, α at the fusion center, i.e, $P_{fa}(\overline{d}_{NP}) = \alpha$ Considering the decisions of the sensors are independent of each other, the fusion center test based on a likelihood ratio test (LRT) can be formulated as follows:

$$L(u) = \frac{P(u|H_1)}{P(u|H_0)} = \prod_{i=1}^{N} \frac{P(u_i|H_1)}{P(u_i|H_0)} = \prod_{i=1}^{N} L(u_i) \tag{4.16}$$

$L(u_i)$ takes two different values, either $\frac{(1-P_{di})}{(1-P_{fai})}$ when $u_i = 0$ with probability $(1 - P_{fai})$ under hypothesis H_0 and probability $(1 - P_{di})$ under hypothesis H_1, or $\frac{P_{di}}{P_{fai}}$ when $u_i = 1$ with probability $(1 - P_{fai})$ under hypothesis H_0 and probability P_{di} under hypothesis H_1. Therefore, we have,

$$L(u) = \frac{\prod_{i=1}^{N} P_{d_i}^{u_i}(1 - P_{d_i})^{(1-u_i)}}{\prod_{i=1}^{N} P_{fa_i}^{u_i}(1 - P_{fa_i})^{(1-u_i)}} \underset{H_0}{\overset{H_1}{\gtrless}} \eta \tag{4.17}$$

Now, employing Equations (4.15)–(4.17), the NP test can be implemented at the fusion center. The person-by-person optimal solution under the independent observation assumption still requires a simultaneous solution of $2^N + N$ equations. This is a point worth noting apparently, the computational complexity of this process grows exponentially with N.

4.10 Bayesian Approach for Spectrum Sensing

A Bayesian detector (BD) is designed for maximization of the spectrum utilization, by digitally modulated primary signals without the prior information on the transmitted sequence of the primary signals. This method makes use of the prior statistics of PU activity and the signaling information of the PU such as symbol rate and modulation order to improve the SU throughput and the overall spectrum utilization of both PUs and SUs. The Bayesian detector is shown having the exact same structure as Neyman-Pearson detector but the design principle of Neyman-Pearson detector is to maximize the detection probability for a given maximal false alarm probability, which results in the difference in detection threshold selection for the two schemes.

There are two hypotheses in spectrum sensing. H_0 is a null hypothesis indicating that PU is absent and H_1 is an alternative hypothesis indicating the existence of PU in a certain spectrum band. There are two important design parameters for spectrum sensing: probability of detection (PD), which is the probability that SU accurately detects the presence of active primary signals, and probability of false alarm (PF), which is the probability that SU falsely detects primary signals when PU is, in fact, absent. We define spectrum utilization as

$$P(H_0)(1 - PF) + P(H_1)PD \tag{4.18}$$

And normalized SU throughput as

$$P(H_0)(1 - PF) \tag{4.19}$$

A point to note is that $P(H_1)PD$ is PU throughput when there are primary signals and the SUs detect the presence of the primary signals. The detection statistics test sample (TS) is compared with a predetermined threshold $\in$ for ascertaining the use of the spectrum by the primary user.

Probability of false alarm PF is the probability that the hypothesis test chooses H_1 while it is in fact H_0:

$$PF = P(TD > \in |H_0). \tag{4.20}$$

Probability of detection PD is the probability that the rest correctly decides H_0 when it is H_1:

$$PD = P(TD > \in |H_1). \tag{4.21}$$

To determine whether the spectrum is being used by the primary user, the detection statistic TD is compared with a predetermined threshold $\in$.

Let us consider time-slotted primary signals where N primary signal samples are used for detecting the existence of primary signals. The PU symbol duration is T which is known to the SU and the received signal $r(t)$ is sampled at $\frac{1}{T}$ at the secondary receiver. The received signal of kth symbol at the CR detector, $r(k)$ is

$$\begin{aligned} &H_0 : r(k) = n(k), \\ &H_1 : r(k) = h(k) * x(k) + n(k), \quad k = 1, 2, 3 \ldots\ldots \end{aligned} \tag{4.22}$$

Assume that the SU receiver has no information on the transmitted signals by the PU and $\varphi n(k)$, $k = 0, 1, \ldots\ldots, N-1$ are independent and identically distributed (i.i.d.) and independent of the Gaussian noise. The detection statistics of energy detector (ED) can be defined as the average energy of observed samples as

$$T_{ED} = \frac{1}{N} \sum_{k=1}^{N} |r(k)|^2 \tag{4.23}$$

The likelihood ratio test (LRT) of the hypotheses H_0 and H_1 can be defined as

$$T_{LRT}(r) = \frac{p(r|H_1)}{p(r|H_0)} \tag{4.24}$$

4.11 Optimal Spectrum Sensing by Using Kullback Leibler Divergence

Sensing performance is measured by two key factors: probability of detection errors and sensing time. A traditional way to design the sensing strategy is based on the Neyman Pearson criterion, while the resulting test fixes the number of required samples, i.e., the sensing time. In the presence of multiple radios, the determination of the optimal threshold value for the test usually involves complex numerical computations. The suitability of a sequential detector is more for collaborative sensing. It is easy to implement and can significantly reduce the average sensing time. In this scheme, each cognitive radio computes the Log-Likelihood ratio for every one of its measurements, while the base station sequentially accumulates the Log-Likelihood statistics and determines the stopping or continuation of making measurements.

4.11.1 System Model

The network consists of M cognitive radios that are monitoring the frequency band of interest, as shown in Figure 4.9. The two hypotheses corresponding to the signal-absent and signal-present events are defined as:

H_0: target signal is absent; H_1: target signal is present.

The signal acquired by the m_{th}, $m = 1, 2, \cdots, M$, cognitive radio device is represented by

$$\begin{aligned} H_0 &: X_m[n] = W_m[n], \\ H_1 &: X_m[n] = S_m[n] + W_m[n], \end{aligned} \tag{4.25}$$

$n = 1, 2, \cdots\cdots$, where $S_m[n]$ the target is signal and $W_m[n]$ is the additive noise. In the approach, the samples acquired by different radios are assumed to be statistically independent, and that the samples acquired by the same radio are independent and identically distributed (i.i.d.). Under H_0 and H_1,

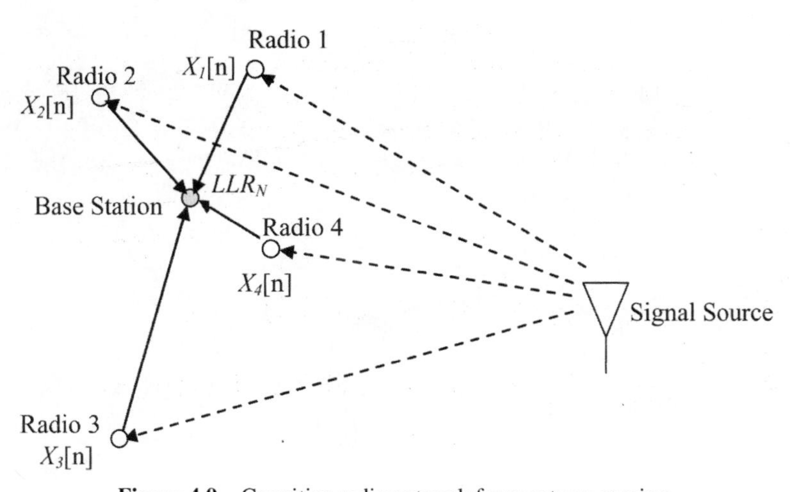

Figure 4.9 Cognitive radio network for spectrum sensing.

the distribution of the acquired signal at the mth, $m = 1, 2, \cdots, M$, radio is characterized by the probability density functions $p_0, m(X_m[n])$ and $p_1, m(X_m[n])$, respectively. The performance of detecting H_0 against H_1 is measured by the probability of false alarm and the probability of miss detection. False alarm refers to the error of accepting H_1 when H_0 is true, while miss detection refers to the error of accepting H_0 when H_1 is true. The probability of false alarm is represented by $PFA = \Pr(H = H_1|H_0)$ and the probability of miss detection is represented by $PMISS = \Pr(H = H_0|H_1)$, where H represents the detector output.

4.11.2 Spectrum Sensing Using KLD

Assume that the number N_{fix} of samples (acquired by each cognitive radio) is *fixed*. The Likelihood ratio test (LRT) is performed at the base station according to

Accept H_1 $if\ LLR > \eta$ for detecting H_0 and H_1

Accept H_0 $if\ LLR \leq \eta$

where the Log-Likelihood ratio LLR is computed as

$$LLR = \ln\left(\prod_{n=1}^{N_{fix}}\prod_{m=1}^{M}\frac{p_{1,m}(X_m[n])}{p_{0,m}(X_m[n]}\right) \tag{4.26}$$

The threshold value η is selected to ensure the probability of false alarm and the probability of miss detection are equal to or less than some pre-assigned values α and β, respectively, i.e.,

$$PFA \leq \alpha,\ PMISS \leq \beta. \tag{4.27}$$

The determination of η usually involves complex numerical computations or simulations. LRT can be implemented for every acquired sample in a sequential manner instead of using a required sample size N_{fix}, for reducing the number of samples required.

Based on (4.27), the expressions for N_{fix} and η are

$$N_{fix} \approx 2\left(\frac{\sqrt{\sum\limits_{m=1}^{M}\left(1-\frac{\sigma_{0,m}^2}{\sigma_{1,m}^2}\right)^2}Q^{-1}(\alpha)+\sqrt{\sum\limits_{m=1}^{M}\left(\frac{\sigma_{1,m}^2}{\sigma_{0,m}^2}-1\right)^2}Q^{-1}(\beta)}{\sum\limits_{m=1}^{M}\frac{\sigma_{1,m}^2}{\sigma_{0,m}^2}-1+\sum\limits_{m=1}^{M}\frac{\sigma_{0,m}^2}{\sigma_{1,m}^2}-2M}\right)^2$$

$$\eta \approx \frac{N_{fix}}{2}\sum_{m=1}^{M}\left(1-\frac{\sigma_{0,m}^2}{\sigma_{1,m}^2}\right)+\frac{N_{fix}}{2}\sum_{m=1}^{M}\ln\left(\frac{\sigma_{0,m}^2}{\sigma_{1,m}^2}\right)+$$
$$+\sqrt{\frac{N_{fix}}{2}\sum_{m=1}^{M}\left(1-\frac{\sigma_{0,m}^2}{\sigma_{1,m}^2}\right)^2} Q^{-1}(\alpha)$$

That is, for $N = 1, 2, \cdots$, we perform the following test:

$$\begin{aligned}&\text{Accept } H_1 \text{ and terminate if } LLR_N \geq A\\&\text{Accept } H_0 \text{ and terminate if } LLR_N \leq B\end{aligned} \tag{4.28}$$

Take one more sample to repeat the test if $B < LLR_N < A$

Where

$$\begin{aligned}LLR_N &= \ln\left(\prod_{n=1}^{N}\prod_{m=1}^{M}\frac{p_{1,m}(X_m[n])}{p_{0,m}(X_m[n])}\right)\\&=\sum_{n=1}^{N}\sum_{m=1}^{M}\ln\left(\frac{p_{1,m}(X_m[n])}{p_{0,m}(X_m[n])}\right)\end{aligned} \tag{4.29}$$

And A, B are predetermined constants according to the sensing objective (2). In the context of cooperative sensing, each radio computes the Log-Likelihood ratio for its every acquired sample, and the base station sequentially accumulates the Log Likelihood statistics and performs the above test, as described in Algorithm 4.1.

A study of how PFA and $PMISS$ depend on A and B is required to understand how A and B are determined. Assume that the detection procedure terminates at $N = N_{stop}$. At $N = N_{stop}$, we have $LLR_{N_{stop}} \approx A$ or $LLR_{N_{stop}} \approx B$, provided that the change in LLR_N at each step is relatively small when compared to the absolute values of A and B. This holds when α and β are sufficiently small.

It can be shown that

$$P_{FA} = \Pr(LLR_{N_{stop}} \geq A|H_0) \approx \frac{1-e^B}{e^A - e^B}, \tag{4.30}$$

$$P_{MISS} = \Pr(LLR_{N_{stop}} \leq B|H_1) \approx \frac{e^{-A}-1}{e^{-A}-e^{-B}} \tag{4.31}$$

Algorithm 4.1 Cooperative sequential detection

0: *Set* $N = 0$.
0: *Set* $LLR0 = 0$ at the base station.
1: *repeat*
2: $N = N + 1$.
3: *The* m^{th}, $m = 1, 2, \cdots, M$, radio acquires sample $X_m[N]$ and computes $\ln(p_{1,m}(X_m[N])/p_0, m(X_m[N]))$.
4: Each radio sends its $\ln(p_{1,m}(X_m[N])/p_0, m(X_m[N]))$ to the base station.
5: The base station updates the Log-Likelihood ratio LLR_N according to

$$LLR_N = LLR_{N-1} + \sum_{m=1}^{M} \ln\left(\frac{p_{1,m}(X_m[N])}{p_{0,m}(X_m[N])}\right)$$

6: **until** $LLR_N \geq A$ or $LLR_N \leq B$.
7: If $LLR_N \geq A$, "H_1 : Target Signal is Present" is claimed; if $LLR_N \leq B$,"H_0 : Target Signal is Absent" is claimed.

For finding appropriate A and B (4.30) and (4.31) are set to be equal to α and β, respectively, and get

$$A = \ln\left(\frac{1-\beta}{\alpha}\right), \quad B = \ln\left(\frac{\beta}{1-\alpha}\right) \tag{4.32}$$

It can be seen A and B do not depend on specific distributions and are convenient to compute.

In the sequential test, the number of samples required to reach a decision is random variable. By using Wald's equation, the average number of required samples under H_0 is given by

$$E_{H_0}\{N_{stop}\} \approx \frac{-(A - B - Ae^B + Be^A)}{(e^A - e^B)\left[\sum_{m=1}^{M} D(p_{0,m}(X_m[n])//p_{1,m}(X_m[n]))\right]} \tag{4.33}$$

Where $D(p(x)//q(x))$ denotes the Kull back-Leibler (KL) Divergence between the distributions $p(x)$ and $q(x)$, i.e.,

$$D(p(x)//q(x)) = \int p(x) \ln\left(\frac{p(x)}{q(x)}\right) dx$$

Under H_1, the average number of required samples is given by

$$E_{H_1}\{N_{stop}\} \approx \frac{A - B - Ae^{-B} + Be^{-A}}{(e^{-A} - e^{-B})\left[\sum\limits_{m=1}^{M} D(p_{1,m}(X_m[n])//p_{0,m}(X_m[n]))\right]} \tag{4.34}$$

Expressions (4.33) and (4.34) show the dependence of average required sensing time on the KL Divergence term provided by each sensing radio. Intuitively, the larger the KL Divergence term, the more the two hypotheses differ from each other, which therefore require less sensing time for detection. In order to save processing power and communication bandwidth, selection of subset of available radios for spectrum sensing may be required.

4.12 Spectrum Sensing Challenges

Several research challenges that need investigation in the area of spectrum sensing function are observed. These are:

- Interference temperature measurement: The problem concerns measurement of interference temperature. Transmit power level and their basic levels are known to an xG user as a matter of course, with the help of a positioning system. However, this ability, and its transmission would mean it is causing interference at a receiver in the vicinity on the same frequency, for which the cognitive radio has no practical ability. Primary receivers are known for their passive nature. So, xG user cannot get the knowledge of the locations of primary receivers. Interference temperature measurement is beyond feasibility.
- Spectrum sensing in multi-user networks: xG networks, as is well known, reside in a multi user environment comprising multiple xG users and primary users. Co-location of xG networks with other xG networks competing for the same spectrum band is also possible. But, current interference models do not take into account the effect of multiple xG users. A multi user environment makes sensing of primary users and estimation of actual interference rather difficult. There is, therefore, need to develop spectrum sensing function considering the possibility of multi-user/network environment. Multiple user problems can be solved through the development of cooperative detection schemes which utilize the spatial diversity found in a multi user network.
- Detection capability: Expeditions detection of primary users is an important requirement for xG networks. OFDM based xG networks are ideal

for the physical architecture of networks and can help reduction in the overall sensing time. There is no need for sensing in other carriers, in the event of detection of a primary user in a single carrier. A proposal for a power based sensing algorithm in OFDM networks for detecting the presence of a primary user is to be proposed. Collection of information helps reduction in over all detection time. But, this process necessitates the use of carriers in large numbers, increasing design complexity in the process. Development of novel spectrum sensing algorithms to ensure minimization of efforts is needed for detection of a primary user within the realization of a given detection error probability.

5

Cognitive Radio for Upper Layers

5.1 Spectrum Management

xG networks are known for the spread of unused spectrum bands over a frequency range that includes both licensed and unlicensed bands. The unused bands detected through spectrum sensing exhibit varying characteristics on the basis of not only the environment that is time varying but also the spectrum band information like operating frequency and bandwidth.

It is imperative for networks to decide on the spectrum band considered best. This is so considering the need to meet QoS requirements over all available spectrum bands. For this reason, new spectrum management functions are required for xG networks, considering the dynamic spectrum characteristics. These functions are categorized as spectrum sensing, spectrum analysis and spectrum decision. The first referred, spectrum sensing, is PHY layer issue while the other two have close relation to the upper layers.

5.1.1 Spectrum Analysis

The available spectrum holes in xG networks exhibit different characteristics varying over time. xG users have the cognitive based physical layer and hence there is the imperative need to understand the features of the different spectrum bands. Spectrum analysis is of great help in understanding such features. Exploitation of these for getting the spectrum band that meets user requirements.

Description of the dynamic nature of xG networks requires characterization of each spectrum on the basis of time-varying environment, primary user activity and information relating to spectrum band including operating frequency and bandwidth. Definition of parameters like interference level, channel error rate, path loss, link layer delay and holding time to reflect the quality of the spectrum is therefore essential. These parameters are briefly described as under

- *Interference*: Some spectrum is highly crowded in comparison with others. It is for this reason that the spectrum band in use determines the interference characteristics of the channel. The permissible power of an xG user can be derived from the quantum of interference at the primary receiver. This information is used for the estimation of channel capacity.
- *Path loss*: There is increase in path loss with increase in operating frequency. Hence, transmission power of an xG remaining the same, its transmission range decreases at higher frequencies with increase in transmission power to make up for increased path loss; there is higher interference for other users.
- *Wireless link errors*: Changes in the error rate of the channel depend on the modulation scheme and the interference level of the spectrum band.
- *Link layer delay*: Addressing different path losses needs wireless link error, interference and varying types of link layer protocols, resulting in link layer packet transmission delay.
- *Holding time*: Channel quality in xG networks is exposed to the activities of primary users. Holding time has reference to the expected time for an xG user to occupy a licensed band before any interruption happens. The longer the duration, the better will the quality be. Spectrum hand off can cause reduction in the holding time. This requires the consideration of previously available statistical patterns of the handoff while designing xG networks with a large expected holding time.

Channel capacity is the most significant factor involved in spectrum characterization. It can be derived from the parameters detailed supra. It is usual for SNR at the receiver to be used for capacity estimation. But, with SNR considering only the observations of xG users, avoidance of interference at primary users is not enough and does not serve any purpose. Spectrum characterization is focused on capacity estimation whose basis is interference at licensed receivers. Interference temperature model can be used for this approach; it provides indication of capacity on the potential RF energy which has the possibility of introduction into the band. It is, therefore, possible to compute or determine the maximum permissible transmission power of xG user with the use of the known quantum of permissible interference.

The spectrum capacity, C can be estimated on this basis as follows:

$$C = B \log\left(1 + \frac{S}{N + 1}\right), \tag{5.1}$$

Where B is the bandwidth, S is the received signal power from the xG user, N is the xG receiver noise power, and I is the interference power received at the xG receiver due to the primary transmitter.

In the backdrop of OFDM based CR system is

$$C = \int_{\Omega} \frac{1}{2} \log_2 \left(1 + \frac{G(f)S_0}{N_0}\right) df \tag{5.2}$$

Where Ω the collection of unused spectrum segments is, $G(f)$ is the channel power gain at frequency f, S_0 and N_0 are the signal and noise power per unit frequency, respectively.

The recent work on spectrum analysis focuses on spectrum capacity estimation. However, other factors that include delay, link error rate, and holding time have also a significant influence on the quality of services. In addition, there is a close relationship of the capacity to both factors, namely, interference level and path loss. An open research issue for enabling identification of the spectrum bands with a combination of all the characterization bands has been referred to supra. This is required for taking a decision on the spectrum that may be considered appropriate for different applications.

5.2 Spectrum Decision

When all available spectrum bands are characterized, selection of appropriate operating spectrum band for the current transmission is imperative considering the QoS requirements and the spectrum characteristics. This requires awareness of user QoS requirements for the spectrum management.

Determination of the data rate, acceptable error rate, delay bound, transmission mode and bandwidth of the transmission is possible on the basis of user requirements. It is then possible to determine the right set of appropriate spectrum bands according to the decision rule. The number of spectrum hand off is used for spectrum decision. This occurs in a specific spectrum band and is done for considering the primary user activity. Spectrum decision happens to be an important issue in xG networks, which have not been explored as yet.

5.2.1 Challenges Faced by Spectrum Management

There is a great need for the investigation of various research needs on the subject of the development of the spectrum decision function.

- ***Design model***: SNR is not adequate for helping characterization of the spectrum band in xG networks. There are many characterization parameters, apart from SNR, that affect quality. These have been referred to earlier. However the way open to combine these spectrum characterization parameters is still an undecided issue. There is a possibility of simultaneous use of multiple spectrum bands in OFDM based xG networks for transmission purpose. Therefore, an imperative need for providing a decision framework for multiple spectrum bands, does exist.
- ***Multiple spectrum band decision***: Simultaneous use of multiple spectrum bands is possible in xG networks for transmission. An xG user can therefore send packets over non-contiguous spectrum bands. Quality degradation is small in multi spectrum transmission during spectrum hand off in comparison with the conventional transmission on a single spectrum band. For instance, when a primary user comes back in a specific spectrum band, the xG user has necessarily to vacate this band. But, abrupt service quality can be reduced as the remaining spectrum bands maintain the communication. There is also the fact that transmission in a multiple spectrum permits the use of lower power in each spectrum band. This results in the achievement of smaller interference with many users. Therefore, there is a great need for spectrum management to support multiple spectrum decision capabilities.
- ***Cooperation with reconfiguration***: An essential feature of cognitive radio is that it enables reconfiguration of transmission parameters of a radio for optimal operation in a specific spectrum band. For instance, when SNR is fixed, adjustment of the bit error rate can be made for maintaining channel capacity through exploitation of adaptive modulation techniques, as, for instance, CDMA 2000, 1xEVDO. Hence, the requirement of a cooperative framework which considers both spectrum decision and reconfiguration.
- ***Spectrum decision over heterogeneous spectrum bands***: As of now, assignment of certain spectrum bands to different purposes has already been completed, while some bands still remain unlicensed. There is, therefore, a likelihood of the spectrum used by xG networks to be a combination of the most exclusively accessed spectrum and an unlicensed spectrum. With licensed bands, xG users need to take into account the activities of primary users in spectrum analysis and decision relating to them, to avoid influencing primary user transmission. As against this, there is a great need for sophisticated spectrum sharing techniques in unlicensed bands, considering all the xG users have equal rights

to spectrum access. The decision on the best spectrum band over this heterogeneous environment requires support from xG network for spectrum decision operations on both the licensed and the unlicensed bands considering the varied characteristics.

5.3 Spectrum Mobility

The use of spectrum in dynamic manners is the target for xG networks. This is done by allowing radio terminals to conduct operations in the best available frequency band. This makes "get the best available channel" possible for the purpose of communication. Realization of the concept requires the capture of the best available spectrum. Spectrum mobility is defined as the process during which an xG user changes its frequency of operation.

5.3.1 Spectrum Handoff

Spectrum handoff arises in xG networks when current channel conditions deteriorate with the appearance of a primary user. Spectrum handoff is the corollary of spectrum mobility giving rise to a new type of handoff. It is imperative for the protocols for different layers of network stack to adapt them to channel parameters of the operating frequency. There should be transparency to the spectrum handoff and the related latency.

A cognitive radio has the adaptability to the frequency of operation. Hence, network protocols shift from one mode of operation to another, every time an xG user changes its frequency of operation. Spectrum mobility management in xG networks has the purpose of ensuring that transitions are smooth and that such applications running on an xG user perceives minimum performance deterioration during a spectrum handoff. It is for the sensing algorithm to provide this information. Learning of this latency puts the mobility management protocols on the job of ensuring smoothness of the transitions and perception deterioration in minimum performance by such applications running on a XG use during a spectrum hand off. As a result, there arises a requirement for multilayer mobility management protocols for accomplishing the spectrum mobility functionalities. Support is rendered by these protocols adaptive to different types of applications. For data communication, as e.g., FTP, there should be implementation of mechanisms from mobility management protocols for storing packets transmitted during a spectrum hand off. But, for a real time application, no need exists for strong packets for the reason that the stored packets will become stale if delivered earlier and cannot be used by the corresponding application.

5.4 Spectrum Mobility Challenges in xG Networks

The open research issues for efficient spectrum mobility in xG networks are discussed below.

- At a specific point of time, several frequency bands may be available for an xG user. Algorithms have to decide the best available spectrum on the basis of the channel characteristics of the available spectrum and the requirements of the applications being used by an xG user.
- Following the selection of the best available spectrum, the next challenge is to design new mobility and connection management approaches for reducing delay and loss during spectrum handoff.
- When the current operational frequency becomes busy (this may happen when a licensed user starts to use this frequency) in the middle of a communication by an xG user, applications run on this node to another available frequency band. But, there is likely take some delay in the selection of a new operational frequency. Novel algorithms are required to ensure applications are not exposed to severe performance degradation during such transitions.
- There is a likelihood of the occurrence of a spectrum hand off for reasons other than the detection of the primary user. Thus, various other spectrum handoff schemes are seen in xG networks. Spectrum band off may occur when a XG user moves from one place to another. Such occurrence may be the result of changes in spectrum bands. There is an imperative need for integration in inter-cell handoff from the desired spectrum handoff scheme. There is also the likelihood of occurrence of spectrum handoff between different networks in xG networks. This is referred to as vertical handoff. A diverse environment of this type makes it necessary for a spectrum handoff scheme to consider all the possibilities mentioned above.
- Spectrum mobility in time domain: xG networks adapt themselves to the wireless spectrum on the basis of available bands on the spectrum. As these available channels change over time enabling QoS in this environment is a challenging job. The physical radio should "move" through the spectrum for meeting the QoS requirements.

- Spectrum mobility in space: There is also a change in as a user moves from one place to another. Allocation of spectrum on a continuous basis is a major challenge in xG networks in this situation.

5.5 Spectrum Sharing

One of the main challenges in xG networks in open spectrum usage is spectrum sharing. This can be regarded as similar to any generic medium access control (MAC) problem in the existing systems. But substantially different challenges are found for spectrum sharing in xG networks. Two factors explain these unique challenges. They are the coexistence with licensed users and the availability of a wide range of spectrum. This section goes into the details of these specific challenges with an overview of the existing solutions and discusses open research areas.

A directory for different challenges during spectrum sharing has to be provided for enumeration of the steps in spectrum sharing in xG networks. This is done first as the challenges. The solutions proposed for these steps are explained in detail. The spectrum sharing process consists of five major steps.

1. ***Spectrum Sensing***: An xG user can allocate only a portion of the spectrum in the event of that portion not being used by an unlicensed user. Challenges involved in the allocation and solutions for them are presented later. So, when the xG works with the objective of transmitting packets, the need arises for awareness of the spectrum usage in its vicinity.
2. ***Spectrum allocation***: The node can allocate a channel on the basis of spectrum availability. There is also the fact that it is determined on the basis of internal policies (and also external polices possibly). This makes the design of a spectrum allocation policy for performance improvement of a node an important topic for research.
3. ***Spectrum access***: Another problem that arises is the need to coordinate the access for preventing collision of multiple users in the overlapping portions of the spectrum, the reason being the likelihood of multiple xG nodes attempting to access the spectrum.
4. ***Transmitter-receiver handshake***: With the determination of a chunk of spectrum for communication, there should also be an indication of the selected spectrum to the receiver for communication. A transmitter-receiver handshake protocol therefore becomes necessary for ensuring efficient communication in xG networks. The handshake does not imply

any restriction for this protocol between the transmitter and the receiver. There is also the possibility of the involvement of a third party like a centralized station.

5. ***Spectrum mobility***: xG nodes are considered as "visitors" to the spectrum allocated. So, when a license user requires specific portions of the spectrum in use, the communication needs continuation in another vacant area. This means that spectrum mobility is important too to ensure success in communication between xG nodes.

The existing work on spectrum sharing in xG networks has the objective of providing solutions for each step referred to supra. The solutions mean a rich literature for spectrum sharing in xG networks.

5.6 Overview of Spectrum Sharing Techniques

The existing solutions for spectrum sharing in xG networks have three aspects: on the basis of their architecture assumption, spectrum allocation behavior, and spectrum access technique. These are shown in Figure 5.1. This section provides a description of these three classifications and present the fundamental results used in the analysis of these classifications. Investigation of the analysis of xG spectrum sharing techniques has been done through use of two major theoretical approaches. While a part of the work uses optimization techniques to find the optimal strategies for spectrum sharing, game theoretical analysis has also found use in this area.

The first classification for spectrum sharing techniques in xG networks is based on their architecture. Details are:

- ***Centralized spectrum sharing***: Here, a centralized entity has control over spectrum allocation and access procedures. With help from these procedures, a distributed sensing procedure is proposed to enable each entity forwarding measurements of the spectrum allocation to the central entity which constructs a spectrum allocation map.
- ***Distributed spectrum sharing***: Distributed solutions are proposed for situations when the construction of an infrastructure finds no preference.

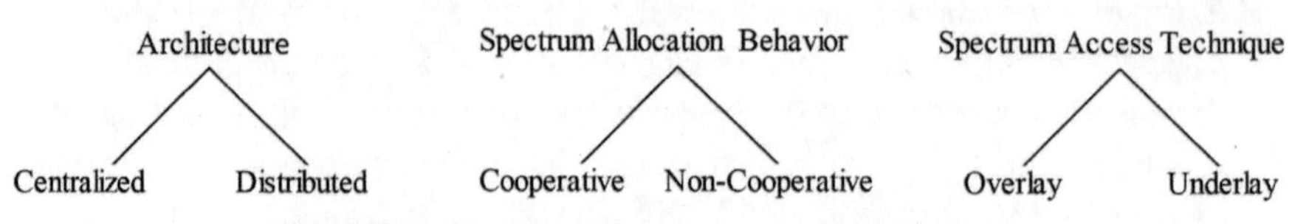

Figure 5.1 Overview of spectrum sharing techniques.

Each node is responsible for the spectrum allocation local (or possibly global) policies form the basis for access.

The second classification of spectrum sharing techniques in xG networks is based on the access behavior. To be precise, the spectrum access can be cooperative or non-cooperative as explained below:

- ***Cooperative spectrum sharing***: Cooperative (or collaborative) solutions consider the effect of the node's communication on other nodes. The interference measurements of each node are shared among other nodes. In addition, the spectrum allocation algorithms consider this information. While all the centralized solutions can be considered as cooperative, distributed cooperative solutions also exist.
- ***Non-cooperative spectrum sharing***: In contrast to the cooperative solutions, non-cooperative (or non-collaborative, selfish) solutions consider only the node at hand. These solutions are also referred to as selfish. While non-cooperative solutions may cause reduced spectrum utilization, the minimal communication requirements among other nodes introduce a trade-off for practical solutions.

Both cooperative and non-cooperative approaches are considered to ensure that cooperative approaches also consider the effect of the channel allocation on the potential neighbors. The simulation results show cooperative approaches outperforming non-cooperative approaches and closely approximating the global optimum. Moreover, a comparison of centralized and distributed solutions reveals distributed solution closely following the centralized solution. Game theory has also been used for performance evaluation of xG spectrum access schemes.

The third classification for spectrum sharing in xG networks is based on the access technology. This is explained below:

- ***Interweave spectrum sharing***: Interweave spectrum sharing refers to the spectrum access technique in which a node accesses the network using a portion of the spectrum not so far used by licensed users. As a result, interference to the primary system is avoided.
- ***Underlay spectrum sharing***: In Underlay spectrum sharing both the primary and the secondary user utilize the same spectrum simultaneously. The secondary user adopts certain techniques like spread spectrum, interference reduction tech; transmit power control etc., to make its signal look like noise at the primary receiver.

- ***Overlay spectrum sharing***: In overlay spectrum sharing, the primary and the secondary user co exist in the same spectrum band. The xG collects the knowledge of the primary user transmission parameters such as code book details, position of the receiver, the OFDM symbol time etc., and uses Orthogonal Codes. beam forming techniques, interference avoidance and transmission of signals at the guard time interval etc.,

The interweave approach becomes more efficient than underlay when interference among users is high. The lack of cooperation among users, however, requires an interweave approach. The comparative evaluation shows the performance loss due to lack of cooperation as small, and disappears with increasing SNR. A hybrid technique is also (spreading based underlay with interference avoidance) investigated where a node spreads its transmission over the entire spectrum and also null or notch frequencies where a primary user is transmitting. Consequently, the interference statistics for each technique are first determined for analysis of outage probability. Then, the outage probability for each technique is then derived on the assumption of the absence of any system knowledge, perfect system knowledge, and limited system knowledge. Similar to other existing works, when perfect system knowledge is assumed, the overlay scheme outperforms the underlay scheme on the points of outage probability. In a more realistic case, the importance of the hybrid technique is exacerbated, when limited system knowledge is considered. The overlay schemes result in poor performance due to imperfections in spectrum sensing. A node can transmit at a channel in an area where a primary user is transmitting. But, the interference caused to the primary user becomes less with the use of underlay with interference avoidance. Another important result is that secondary users of a large number can be accommodated by the hybrid scheme compared to the pure interference avoidance scheme.

5.7 Inter-network Spectrum Sharing

xG networks have been introduced for providing access opportunities to the licensed spectrum for unlicensed users. Deployment of multiple systems in overlapping locations and spectrum is made possible. This is shown in Figure 5.2. This makes spectrum sharing a topic of significance for research in networks. Regulation of inter-network spectrum has so far been regulated through static frequency assignment among different systems. Ad-hoc systems have seen only investigation of the interference issues in ISM band. This is on most occasions of the co-existence of WLAN and Bluetooth networks. As a

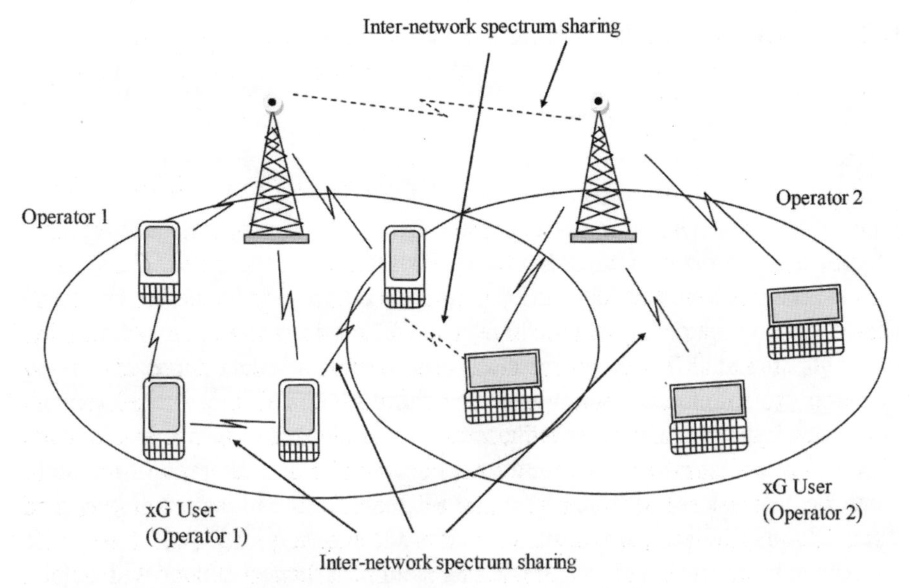

Figure 5.2 Inter-network spectrum sharing.

result, challenges of a unique type are seen in inter-network spectrum sharing in xG networks.

5.8 Centralized Inter-Network Spectrum Sharing

The common spectrum coordination channel (CSCC) etiquette protocol is proposed for the coexistence of IEEE 802.11 b and 802.16 a networks. This is the first step seen for the co-existence of open spectrum systems. But this is not considered as a solution of xG network, because it makes modifications in users of both networks as a necessary factor. There is the assumption that each node should be equipped with a cognitive radio and a low bit rate, narrow band control radio. Coordination of these nodes through broadcasting of CSCC messages helps maintaining the co-existence. The channel for use is determined by the user in each case for facilitating avoidance of interference. In the event of inadequacy for avoiding interference, deployment of power adaptation is resorted to. The evaluations bring to light the improvement in CSCC protocol throughput by 35–160% via both frequency and adaptation when there is a vacant spectrum for using frequency adaptation.

5.9 Distributed Inter-Network Spectrum Sharing

The basic concept forming the basis for distributed QoS based dynamic channel reservation (D-QDCR) is that a base station (BS) of wireless internet service providers (WISPs) competes with its interference BSs as per the QoS requirements of its users for allocation of a portion of the spectrum. The control and data channels get separated in the same way as the common spectrum coordination channel (CSCC) protocol. Q frame is the basic unit for channel allocation in distributed QoS based dynamic channel reservation (D-QDCR). Q-frame, when is allocated by BS, it uses the control and data elements allocated for the purpose of coordination and data communication between users competition between BSs takes place on the priority of each BS depending on BS data volume and QoS requirement. There is also the proposal different type of competition policies based on the type of transfer from the user's demand. D-QDCR scheme play the role of a contribution for inter-network sharing.

The inter-network spectrum sharing solutions have been providing an expansive view of spectrum sharing solution that includes some operator policies to enable determination of the spectrum allocation. The existing solutions in xG network architecture face a major problem which is the requirement of a common control channel.

5.10 Challenges to Spectrum Sharing

Earlier sections have dealt with the theoretical findings and solutions for spectrum sharing in xG networks. Despite the existence of a large number of research findings, there is still a need for research on the topic of efficient and seamless open spectrum operation.

5.10.1 Common Control Channel (CCC)

Many spectrum sharing solutions,- centralized or distributed – are on the basis of the assumption of a CCC for spectrum sharing. CCC facilitates many spectrum sharing personalities like a transmitter-receiver handshake, communication with a central entity, or a sensing information exchange. As mentioned earlier, xG network users are considered 'visitors' to the spectrum allocated by them. This means that the channel has to be vacated by the primary user when he uses it. This holds good for the CCC too. This leads to infeasibility in implementing a fixed CCC in xG networks. In addition, a channel common for all users in a network with primary users is shown to be dependent much on the relevant topology and, as a result, with variation over time.

5.10.2 Dynamic Radio Range

Changes in radio range take place with operating frequency arising from variation in alternation. There is the assumption of a fixed range in many solutions independent of the operating spectrum. But, there is likelihood of neighbors of a node changing with changes in operating frequency, in any case where a large portion of wireless spectrum is considered. This has an effect on the interference profile and routing decisions. Further, the choice of a control channel needs a careful decision as a consequence of this property. Selection of a control channel in the lower portions where the transmission range will be higher is desirable and efficient. To this may be added the selection of data channels in the higher portions of the spectrum with the possibility of utilizing a localized operation with minimized interference. Till date, there has been no work dealing with this important challenge in xG networks. So operation frequency aware spectrum sharing techniques are recommended, considering the direct inter-dependency existing between interference and radio range.

5.10.3 Spectrum Unit

Most of the spectrum sharing techniques regards a channel as the basic spectrum unit for operation in xG networks. Many algorithms and methods have been suggested for selecting the suitable channel to ensure efficient operation in xG networks. But, there have been ambiguous definitions of a channel as "Orthogonal non-interfering", TDMA, FDMA, CDMA or a combination of these" in many places or as "a physical channel as in IEEE 802.11" or "a logical channel associated with a spectrum region or a radio technology". In some places, the channel goes by the definition of the frequency as a frequency bands. Apart all these, the fact remains that a clear and unambiguous definition is crucial for further development of algorithm. The characteristic features of a channel may not remain the same all through in view of its effects on operating efficiency. In sharp contrast, there is the assumption of a channel providing the same bandwidth as other channels. In addition to this, primary users and the feature of heterogeneity present in the networks bring in additional challenges for the choice of a spectrum unit/channel. This leads to the suggestion of allocation of different units such as CSMA random access, TDMA time slots, CDMA codes and also hybrid types to primary users. Provision of seamless operations requires consideration of these properties in the choice of a spectrum unit.

5.11 Upper Layer Issues

xG networks face challenges in terms of spectrum sensing, spectrum management, spectrum mobility, spectrum sharing, upper layer issues such as routing, flow control and congestion control – all are important for the realization of spectrum in xG networks. The challenges related to these areas are discussed now.

5.11.1 Routing Challenges

Routing is an important but yet unexplored problem in xG networks. The exclusive characteristics of open spectrum phenomenon require development of innovative routing algorithms for xG networks which have multi-hop communication requirements. Till date, the focus of research on xG networks has been on spectrum sensing techniques and spectrum sharing solutions. But the emphasis has now shifted to the need for routing algorithms in an open spectrum environment which ranks as an important topic in xG network. Collaboration between routing and spectrum management is a major design choice for routing in xG networks. This need arises considering the intermittent nature of the dynamic spectrum as chronological and spatial. Simulation based comparisons are done for a cross layer solution that makes routes and determines the operating spectrum in combination for each hop that outperforms a sequential approach where selection of routes is done without depending on spectrum allocation.

There is also the proposal of a cross-layer solution which involves a combination of route selection and spectrum management. In this scheme, DSR is used by each node for locating candidate paths, scheduling a time and channel for each hop. This obviously is a source-based routing technique centrally performed with the use of a global view of the network to exhibit the upper bound in achievable performance. One more exclusive challenge for xG networks routing lies in the development of tools for analytical evaluation of routing protocols. Analysis of routing protocols for ad hoc networks with the use of graph models is the traditional practice. But, these networks have the limitation that communication spectrum is fixed and continuous in sharp contrast to the dynamic nature of xG networks. There is, therefore, the possibility of a node using the same set of static channels for communication with neighbors.

Another unique challenge for routing in xG networks is the development tools for analytical evaluation of routing protocols. Traditionally, analysis of routing protocols for ADHOC networks is done using graph models. However, in these networks, the communication spectrum is fixed and continuous contrary to the dynamic nature of xG networks. Hence, a node can use the same set of static channel(s) for communication with all neighbors. Against this, modeling network topology and connectivity of an xG network is a challenging job. Open research issues for routing in xG networks are:

- ***Common control channel***: The lack of a common control channel (CCC) in xG networks constitutes a major problem. This has been discussed earlier. Traditional routing protocols require either local or global broadcast messages for specific functionalities such as neighbour discovery, route discovery and route establishment. But, even broadcasting in xG networks is a problem in view of the lack of a CCC. Hence, solutions considering this fact are required in xG networks.
- ***Intermittent connectivity***: In xG networks, there is the likelihood of a rapid change in the neighbours of a node who can be reached. Two factors explain this phenonomenon. The first is the change or disappearance of the available spectrum licensed users exploit the network. The reachability ceases to exist once a node selects a channel for communication, with the result that the connectivity concept used for wireless networks is different in xG networks while depending on the spectrum.
- ***Re-routing***: In xG networks, the intermit-tent connectivity may cause a change in the route established for a flow from the available spectrum in addition to mobility. Hence, re-routing algorithms that consider the dynamic spectrum are necessary for routing in xG networks.
- ***Queue management***: Queue management is one more challenge in xG networks not addressed so far. There is the possibility of an xG terminal having multiple interfaces for communication with different nodes. These interfaces may cease to be available due to variations in the available spectrum over time and require the packets served through any interface moving to some other interface. Deployment of various priorities of different types to by the requirements of the QoS. This leads to the need for investigation of the implementation of a single queue or multiple queues for each traffic type of each interface.

5.12 Transport Layer Challenges

Transport protocols constitute an unexplored area for xG networks. Several solutions have been proposed for improvement in the performance of TCP and UDP in conventional wireless networks in recent years. These studies focus on mechanisms to limit the performance degradation of TCP and UDP arising as a result of wireless link errors and access delays. However, the xG networks bring in new challenges for transport protocols.

The performance of TCP depends on the packet loss probability and the round trip time (RTT). Wireless link errors and, consequently, the packet loss probability depend not only on the access technology, but also on the frequency in use, interference level and the available bandwidth. Therefore, the wireless TCP and UDP protocols designed for the existing wireless access technologies will not serve the purpose in dynamic spectrum assignment based xG networks.

As against this, RTT of a TCP connection has indirect dependence on operation frequency. For instance, there is the requirement of a larger number link layer retransmissions for successful transport of a packet across a wireless channel. This occurs when the packet error rate (also referred to as the frame error rate) is higher at a particular frequency band. In addition, wireless channel access delay seen in xG networks depends on operation frequency, interference level and medium access control protocol. RTT of a TCP connection is influenced by these factors. Hence, there is variation in the probability of RTT while packet loss observed by a TCP protocol may occur on the basis of the frequency of operation.

There may be variation in the operation frequency of a cognitive radio in xG networks from time to time due to spectrum handoff. A change in xG terminal changes its operating frequency, resulting in a finite duration of delay before the new frequency can be operational. This is referred to as the spectrum handoff latency. The spectrum handoff latency can increase the RTT, leading to retransmission timeout (RTO). Conventional transport protocols can see this RTO as packet loss and call for its congestion avoidance mechanism resulting in reduced throughput. Transport protocols need to be designed to enable their transparency spectrum handoff. Design of transport protocols is required for eliminating the adverse effects of spectrum mobility. These should be transparent to spectrum handoff.

Design of innovative communication protocols in xG networks has become necessary as a result of challenges in communication. Properties of the spectrum band in use influence the performance of xG networking functionalities, which in turn brings in the necessity for a cross layer design in the entire gamut of the networking protocol stack. Careful attention has to be paid to the design of communication protocols to the effects of the selected band and changes arising out of spectrum mobility. There is also the imperative need for spectrum management functionalities like spectrum sensing and spectrum handoff to collaborate with communication protocols.

5.13 Cross-Layer Challenges in Spectrum Management

Adaptation of communication protocols to wireless channel parameters becomes necessary as a result of the dynamic nature of the underlying spectrum in xG networks. There is also the fact that the behavior of each protocol affects the performance of other protocols. To quote an instance, medium access techniques of various hues that find use in xG networks have a direct effect on the round trip time (RTT) for the transport protocols. In the same way, there is a change in RTT and error probability where re-routing is done due to link failures arising from spectrum mobility. Such a change in error probability affects the overall performance of the medium access protocols and, as consequence; all these changes affect the overall quality of the user applications.

Cooperation exists between spectrum management and communication layers. Decision on the appropriate spectrum band leads the spectrum band to the requirement of information on QoS requirement transport, routing, scheduling and sensing. These interdependencies among the functionalities of communication stack and their closely getting coupled with the physical layer give rise to the imperative need for a cross layer spectrum management function that takes note of medium access, routing and transport, application requirements and also the spectrum available in the selection of the operating spectrum.

5.14 Cross-Layer Challenges in Spectrum Handoff

Spectrum handoff results in latency, affecting the performance of the communication protocols. The main challenge in spectrum handoff is the reduction in the latency for spectrum sensing. The spectrum handoff latency has deleterious effects on the performance of transport protocols. During spectrum handoff, channel parameters such as path loss, interference, wireless link error, and link layer delay are influenced by the dynamic use of the spectrum. As against this, changes in the PHY and MAC channel parameters can trigger spectrum handoff. The user application may request spectrum handoff to find a better quality spectrum band.

The spectrum mobility function works hand in hand in close cooperation with the spectrum management function and spectrum sensing to decide on an available spectrum band. Estimation of the effect of spectrum handoff latency requires information on the link layer and sensing layers. There should also be awareness of the latency for reducing any abrupt quality degradation on the part of transport and application layer. Routing information is important for route recovery through use of spectrum handoff. These factors bring in close relation of spectrum handoff to the operations in all communication layers.

5.15 Cross-Layer Challenges in Spectrum Sharing

The performance of spectrum sharing has a direct dependence on the spectrum sensing capabilities of the xG users. Spectrum sensing is basically a PHY layer function. However, with cooperative detection, there is imperative need for xG users to use ADHOC connection for exchanging sensing information, which necessitates a cross-layer design between spectrum sharing and spectrum sensing. The performance of communication protocols depends on spectrum sensing, i.e., getting information about the spectrum utilization. Two major challenges are seen in this aspect.

Interference mitigation is the first challenge. When interference comes in at the receiver, spectrum sensing alone provides information on transmitters. This leads to the requirement of transmitters for considering both their interference to other users and interference at their receivers. Cooperative techniques are known for their superiority in terms of system performance. As against this, such collaboration causes increase in communication overhead. This may cause degradation in system performance degradation when channel capacity or energy consumption is considered. As a result, effective spectrum sharing techniques that make efficient collaboration possible between different xG nodes in terms of spectrum sensing information sharing a requirement.

The second challenge for spectrum sensing is the inability to get the entire spectrum sensed all the time. There is the requirement of a big chunk of time needed to sense the entire spectrum, arising out a big range of spectrum foreseen for xG networks. This process calls for careful scheduling considering sensing needs and energy. So the assumption that a node has accurate knowledge of the spectrum on all occasions is not practical. Current radio technologies stand in the way of continuous spectrum sensing when a single radio is deployed on a device. Spectrum sensing has to be stopped during communication to enable switching to the required channel and do communication. This is the reason why this operation requires cross layer interaction between the physical and the upper layers. There is attempt by communication to get coordinated with spectrum sensing events. The alternative for this is the investigation of the effect of using multiple radios, where a two transceiver operation is considered to enable transceiver listening to the control channel for sensing. The operation improves system performance. However the complexity and device costs are high.

The spectrum sharing techniques discussed earlier decouples spectrum allocation and spectrum sensing. In the decentralized cognitive MAC (DC-MAC) scheme proposed earlier, a cross-layer approach for spectrum allocation and spectrum sensing is considered. Spectrum sensing is in conjunction with spectrum allocation and application layer such that the node sensing only a portion of the spectrum and allocating channel in response to a request from the application layer. First, an optimal approach for channel allocation is provided followed by a greedy suboptimal approach. The ability of a node to serve a channel among N channels is assured. Performance evaluation shows the greedy approach closely approximating the optimal solution. This solution is also robust to inaccuracies in spectrum sensing and limited knowledge.

5.16 Cross-Layer Challenges in Upper Layers

Considering available spectrum bands in xG networks with multi-hop communication are different between gaps, spectrum sensing information is required for topology configuration in xG networks. A major design choice for routing in xG net-works is the collaboration between routing and spectrum decision. When the optimal route from an xG user to another results in an interference to the primary users, the end-to-end latency or packet losses can get affected along the route consequently on this interaction. Mitigation of degradation can be done by selecting multiple spectrum interfaces at intermediate nodes. Therefore, end-to-end route may consist of multiple hops traversing different spectrum bands.

Finally, re-routing needs performance in a cross-layer fashion. If a link failure occurs as a result of spectrum mobility, the routing algorithm needs to distinguish this failure from a node failure. Moreover, intermediate xG users can perform re-routing by exploiting the spectrum information available from spectrum sensing functionality to select better routes method.

Variations are seen in RTT and packet loss rates as a result of the dynamic frequency of operation. This in turn leads to variations in packet transmission delay. Moreover, medium access schemes cause access delay. All these factors influence the round trip time of a connection, which, in turn, affects the performance of transport protocols. The latency associated with a spectrum handoff increases the instantaneous RTT of the packet transmission. As a result, transport protocols for xG networks require designing in a spectrum aware approach that introduces cooperative operation with the other communication layers.

5.17 MIMO Cognitive Radio

Wireless transmissions via MIMO transmissions have received considerable attention during the past decade. MIMO technology serves serving as a building block of next-generation wireless communication systems supporting much higher data rates than UMTS and HSDPA based 3G networks. MIMO is actually a signal processing technique used to increase the performance of wireless communication systems using multiple antennas at the transmitter and receiver. A general MIMO system model is shown in Figure 5.3. A communication system with N_T transmit antennas and N_R receive antennas is presented. Antennas $T_{x1} \ldots\ldots T_{xNT}$ respectively send signals $x_1 \ldots\ldots x_{NT}$ to receive antennas $R_{x1} \ldots\ldots R_{xNR}$. Each receiver antenna combines the incoming signals which coherently add up. The received signals at antennas $R_{x1} \ldots\ldots R_{xNR}$ are respectively denoted by $y_1 \ldots\ldots y_{NR}$. We express the received signal at antenna $T_{xq}; q = 1, \ldots\ldots, NR$ as:

$$y_q = \sum_{p=1}^{N_T} h_{qp} \cdot x_p + b_q \qquad where\ q = 1, \ldots\ldots\ldots N_R$$

The flat fading MIMO channel model is described by the input-output relationship as given below

$$y = H \cdot x + b$$

where H is the $(N_R \times N_T)$ complex channel matrix given by:

$$H = \begin{pmatrix} h_{11} & h_{12} & \cdots & h_{1_{NT}} \\ \vdots & \vdots & \ddots & \vdots \\ h_{N_R 1} & h_{N_R 2} & \cdots & h_{N_R N_T} \end{pmatrix} \tag{5.3}$$

h_{qp} is the complex channel gain which links transmit antenna T_{xp} to receive antenna R_{xq}.

Here $x = x_1, \ldots\ldots\ldots, x_{NT}$ T is the $(N_T \times 1)$ complex transmitted signal vector.

$y = y_1, \ldots\ldots\ldots, y_{NR}$ is the $(N_R \times 1)$ complex received signal vector.

$b = b_1, \ldots\ldots\ldots, b_{NR}$ is the $(N_R \times 1)$ complex additive noise signal vector. The continuous time delay MIMO channel model of the $(N_R \times N_T)$ MIMO channel H associated with time delay τ and noise signal $b(t)$ is expressed as:

$$y(t) = H(t),\ \tau x(t - x)d\tau + b(t) \tag{5.4}$$

where $y(t)$ is the spatial temporal output signal, $x(t)$ is the spatial temporal input signal, $b(t)$ is the spatiotemporal noise signal.

MIMO technology helps in achieving the performance of many desirable functions for wireless transmissions, such as folded capacity increase without bandwidth expansion, dramatic enhancement of transmission reliability via space-time coding and effective co-channel interference suppression for multiuser transmissions. Such advantages have triggered the adoption of MIMO in the next generation WiFi, WiMax and cellular network standards.

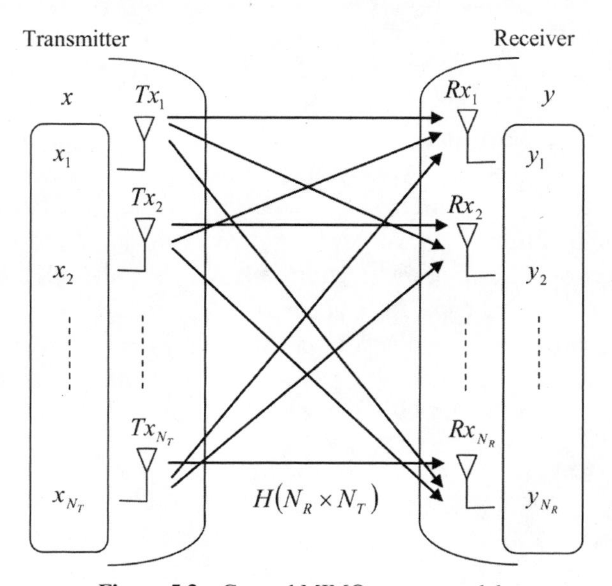

Figure 5.3 General MIMO system model.

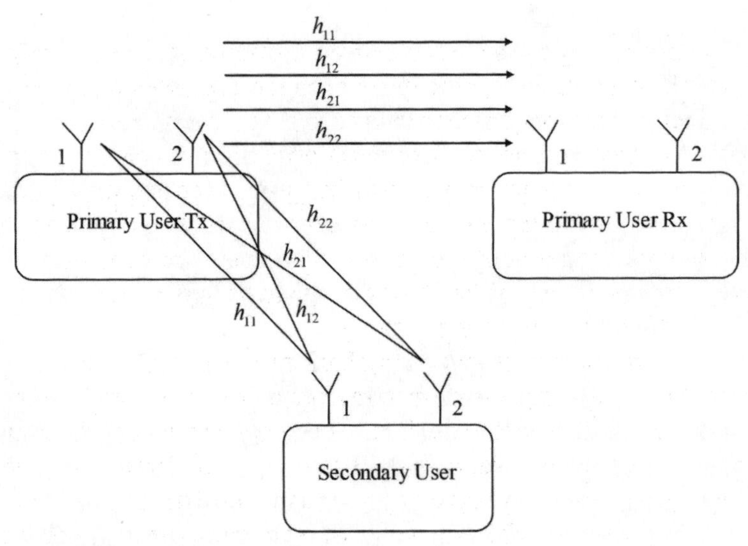

Figure 5.4 Basic two-antenna MIMO system incorporated in CR.

MIMO technology proves useful also in spectrum sensing in Cognitive Radio. Most prior research on radio resource allocation for CR networks has assumed a single antenna at both primary and secondary transceivers. But here, multi-antennas are placed both on the PU and on SU. e.g. two antenna systems as shown in Figure 5.4. These multiple antennas can be used for allocating transmit dimensions in space and thereby providing the secondary user higher degrees of freedom in space, in addition to time and frequency; so as to balance between maximizing its own transmits rate and minimizing the interference powers at the primary users. Transmission of signals through different paths, exploiting receiver and transmitter diversity, maintains reliable communication between PU and SU. Thus, there is an increase in both capacity and spectral efficiency. MIMO technique helps in combating multipath scattering between PU and SU by providing spatial diversity leading to reliable sensing. Further, this technique exploits multipath scattering by providing spatial multiplexing leading to higher throughput.

6

Standards for Cognitive Radio IEEE 802.22 Wireless Regional Area Network

6.1 Introduction

Cognitive radio has recently emerged as a useful technology with ability to improve the efficiency of spectrum utilization. As a traditional practice, the spectrum is assigned by the Federal Communications Commission (FCC) to specific users or applications, and each user can only utilize its pre-assigned bandwidth for communication. This discipline causes some overcrowding in bandwidth while some other bandwidth may be seriously under-utilized. The concept of cognitive radio aims at providing a flexible way of spectrum management, permitting secondary users to temporally access spectrum that is not currently used by legacy users. In this regard, the FCC has taken a number of steps towards allowing low-power devices to operate in the broadcast TV bands that are not being used by TV channels. The TV bands include the following portions of the VHF and UHF radio spectrum: 54–72 MHz, 76–88 MHz, 174–216 MHz and 470–806 MHz Each TV channel occupies a slot of 6 MHz bandwidth. When a TV frequency band is not used in a particular geographical region, it can be used by cognitive radios for transmission. IEEE has established the IEEE 802.22 Working Group for promoting this development and developing in the vicinity standard for a cognitive radio-based device for TV bands.

A key challenge in the development of the IEEE 802.22 standard is that a cognitive radio should have the ability to reliably detect the presence of TV signals in a fading environment. This work was supported in part by the National Science Foundation under grants ECS-0601266 and ECS-0725441. Otherwise, the radio may use the frequency band occupied by a TV channel, and consequently cause serious interference to the TV receivers in the vicinity. Many sensing or detection schemes have been recently reported in the IEEE 802.22 community. These schemes can be classified into two categories:

single-user sensing and cooperative sensing. Large variations in the received signal strength caused by path loss and fading, single-user sensing have demonstrated the unbelievable feature of which consequently triggered the FCC to require geolocation-based methods for identifying unused frequency bands. The geolocation approach is suitable for registered TV bands. However, its cost and operational overhead prevent its wide use in the opportunistic access to occasional "white spaces" in the spectrum. Cooperative sensing relies on multiple radios in the detection of the presence of primary users, and provides a reliable solution for cognitive radio networks.

6.2 IEEE 802.22 Wireless Regional Area Network

The coverage area of IEEE 802.22 cell size ranges from 33 to 100 km and consisting of point to multipoint (P-MP) network comprising a base station and the customer premise equipments (CPE). WRAN cell can provide services to 512 users. The downlink rate is about 1.5 Mbps per CPE and uplink rate of 384 Kbps. FCC selected the TV bands for providing such a service because these frequencies feature propagation characteristics, considered favorable and which would allow far out users to be serviced and hence provide a suitable business case for Wireless Internet Service Providers (WISPs).

6.2.1 Importance of IEEE 802.22

In developing countries, particularly India and China, more than 70% of the population lives in rural areas. Providing satisfactory communication services to these people is considered as one of the key factors to bring them on to par with the urban population in terms of economic and social development. The rural areas are geographically widespread, making provision of broadband service at affordable cost a difficult and challenging goal. Installation of wired lines is not a solution for the above problem. Hence, with the advent of wireless technology, providing communication services to rural areas is likely to become viable.

Today, wireless technologies such as cellular, satellite and Wi-Fi are most suited for providing Internet facilities in rural areas. However, there is a perennial need for low-cost services in rural areas. A point to note is that the licensed spectrum below 3 GHz has significant unused capacity at any given space and time. The economic potential for the TV white spaces has been estimated at $-100 billion. In 2004, the FCC started investigating underutilized TV bands for two-way communication. Later the FCC allowed unlicensed use of unused TV band spectrum on November 42008. On February 17, 2009; the FCC released certain rules for unlicensed operation in the TV broadcast bands. IEEE 802.22 promises to be one of the major developments towards making communication services accessible in the rural areas.

6.2.2 Topology of IEEE 802.22

The 802.22 system specifies a fixed P-MP wireless air interface whereby a BS manages its own cell and all associated CPEs. This is depicted in Figure 6.1. The BS (a professionally installed entity) controls the medium access in its cell and transmits in the downstream direction to the various CPEs, which respond back to the BS in the upstream direction. In addition to the traditional role of a BS, it also manages a unique feature of distributed sensing.

This is needed to ensure proper incumbent protection and is managed by the BS, which instructs the various CPEs to perform distributed measurement of different TV channels.

6.2.3 Service Capacity and Coverage

The IEEE 802.22 system specifies spectral efficiencies in the range of 0.5 bit/sec/Hz up to 5 bit/sec/Hz. Consideration of an average of 3 bits/sec/Hz would correspond to a total physical layer data rate of 18 Mbps in 6 MHz

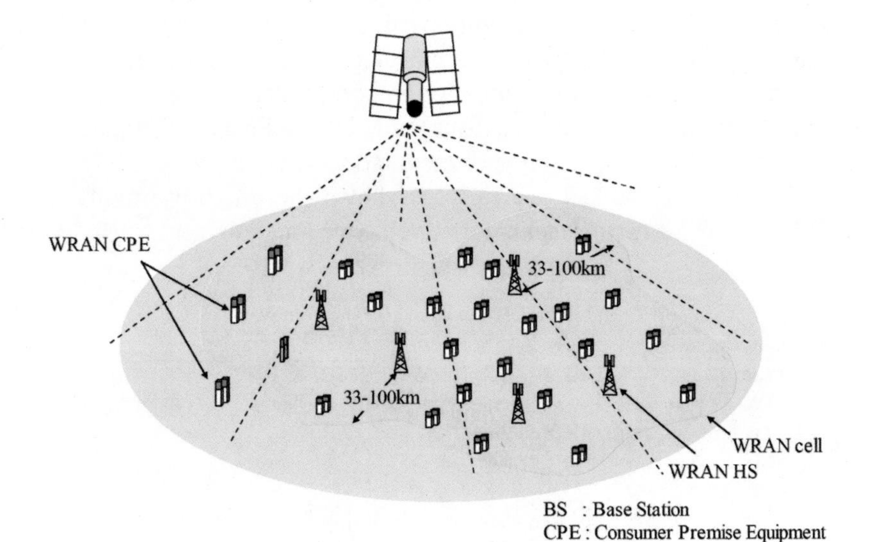

Figure 6.1 Topology of IEEE 802.22.

TV channel. In order to obtain the minimum data rate per CPE, a total of 12 simultaneous users have been considered which leads to the required minimum peak throughput rate at edge of coverage of 1.5 Mbps per CPE in the downstream direction. A peak throughput of 384 kbps is specified, in the upstream direction which is comparable to digital subscriber loop (DSL) services. Another distinctive feature of IEEE 802.22 WRAN as compared to existing IEEE 802 standards is the BS coverage range, which can go up to 100 Km if power is not an issue (current specified coverage range is 33 Km at 4 Watts CPE EIRP). WRANs have a much larger coverage range than today's networks, which is primarily due to its higher power and the favorable propagation characteristics of TV frequency bands. Table 6.1 summarizes the simulation parameter.

6.3 Physical Layer

In the specific case of the PHY, there is need to offer high performance while keeping the complexity low. In addition, there is need for efficient exploration of the available frequency to provide adequate performance, coverage and data rate requirements of the service. WRAN applications require flexibility on the downstream with support for variable number of users with possibly variable throughput. WRANs have also a need to support multiple access on the upstream. Multi-carrier modulation is very flexible in this regard, as it enables control over the signal in both time and frequency domains. This provides an opportunity to define two-dimensional (time and frequency) slots and to map the services for transmission in both directions onto a subset of these slots. Details of TV band occupancy are given in Figure 6.2 802.22 PHY to provide high flexibility in terms of modulation and coding.

Preliminary link budget analysis has shown the difficulty involved in meeting the 802.22 requirements (about 19 Mbps at 30 Km) by using just

Table 6.1 IEEE 802.22 standard specification parameters

Parameters	Specifications
Typical cell radius (km)	30–100 km
Methodology	Spectrum sensing to identify free channels
Channel bandwidth (MHz)	6, (7, 8)
Modulation	OFDM
Channel capacity	18 Mbps
User capacity	Downlink: 1.5 Mbps Uplink: 384 kbps

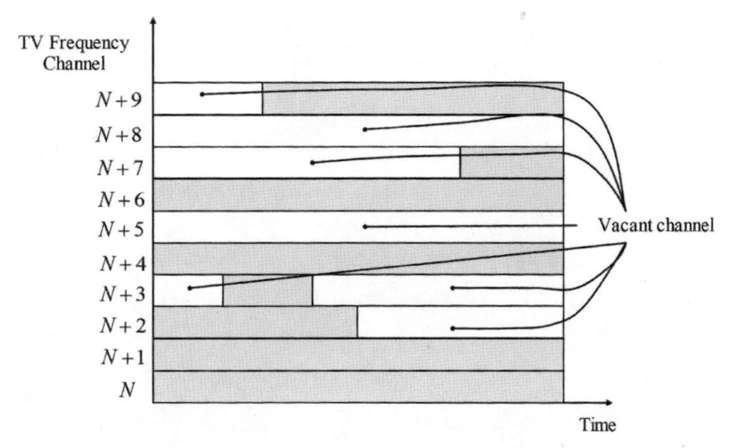

Figure 6.2 TV band occupancy over time and frequency.

1 TV channel for transmission. The use of channel bonding by aggregating contiguous channels enables meeting this requirement allows this requirement to be met. There are two channel bonding schemes.

- Contiguous bonding
- Non-contiguous bonding

Figure 6.3 is the simplified diagram of the contiguous channel bonding scheme. In principle, bonding as many TV channels as possible is desirable. However, limitations seen in the practical implementation impose constraints on the number of channels that can be bonded. For implementation purposes, it is desirable to limit the bandwidth of the RF front-end part of the communication system. The current US grade-A TV allocation restricts adjacent allocated TV channels to have at least 2 empty channels between them. This is done so to reduce interference from one high-power TV channel to the other. Thus, the minimum vacant TV channel spacing needed for the WRAN device to operate is 3 TV channels. Based on this, RF bandwidth is limited to 3 contiguous channels. This implies a RF bandwidth of 18 MHz for 6 MHz TV channels.

6.4 MAC Layer

The CR-based MAC needs to be highly dynamic for quick responses to changes in the operating environment. Besides providing traditional MAC services, the 802.22 MAC is required to perform an entirely new set of functions for effective operation in the shared TV bands.

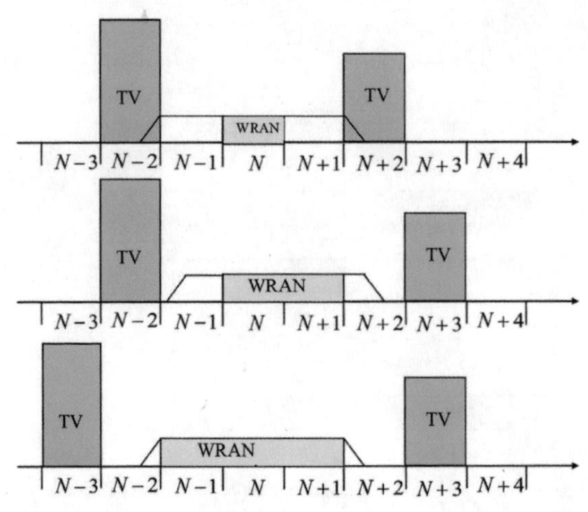

Figure 6.3 Channel bonding scheme illustrating1 (top), two (middle), and three TV channels (bottom).

6.4.1 Super Frame Structure

The current 802.22 draft MAC employs the super frame structure depicted in Figure 6.4. At the beginning of every super frame, the BS sends a special preamble and SCH (super frame control header) through each and every TV channel (up to 3 contiguous) that can be used for communication and that is guaranteed to meet the incumbent protection requirements. CPEs are tuned to any of these channels and those who synchronize and receive the SCH, are able to obtain all the information it needs for association with the BS.

During the lifetime of a super frame, multiple MAC frames are transmitted, these may span multiple channels and hence can provide better

- System capacity
- Range
- Data rate
- Multipath diversity

However, for the purpose of flexibility, the MAC supports CPEs which are capable of operating on a single or multiple channels. During each MAC frame the BS has the responsibility to manage the upstream and downstream direction, which may include ordinary data communication, measurement activities and coexistence procedures.

In the 802.22 draft MAC, whenever a CPE starts up, it first scans (perhaps all) the TV channels and builds a spectrum occupancy map that helps ascertainment of the detection or otherwise for each channel. This information may be

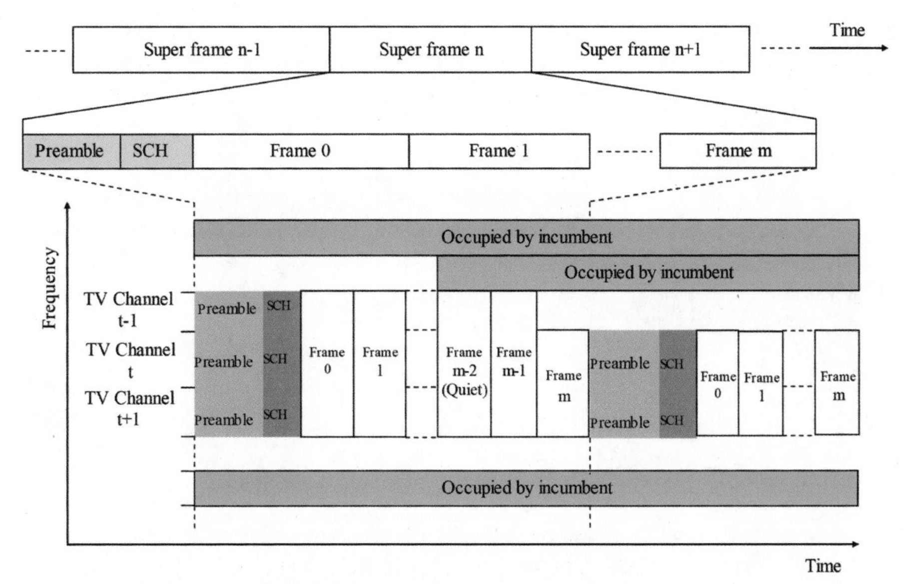

Figure 6.4 Super frame structures.

conveyed later to a BS and also used by the CPE to determine which channels are vacant and hence use them to look for BSs. In those vacant channels, the CPE must then scan for SCH transmissions from a BS. The duration a CPE stay in a channel is equal to the super frame duration. Once the CPE receives the SCH, it acquires channel and network information that is used to proceed with network entry and initialization. The MAC frame structure is shown in Figure 6.5. A frame comprises of two parts: a downstream (DS) sub frame and an upstream (US) sub-frame. The boundary between these two segments is adaptive, thereby facilitating the control of the downstream and upstream capacity. The downstream sub frame consists of only one downstream PHY PDU with possible contention intervals for coexistence purposes. An upstream sub-frame consists of contention intervals scheduled for initialization (e.g., initial ranging), bandwidth request, UCS (Urgent Coexistence Situation) notification, and possibly coexistence purposes and one or multiple upstream PHY PDUs, each transmitted from different CPEs.

6.5 Sensing in IEEE 802.22

One of the key elements of the 802.22 standard is the fact of its coexistence it can coexist with other users of the radio spectrum, without causing any interference. As any 802.22 system is likely to be given access to any spectrum

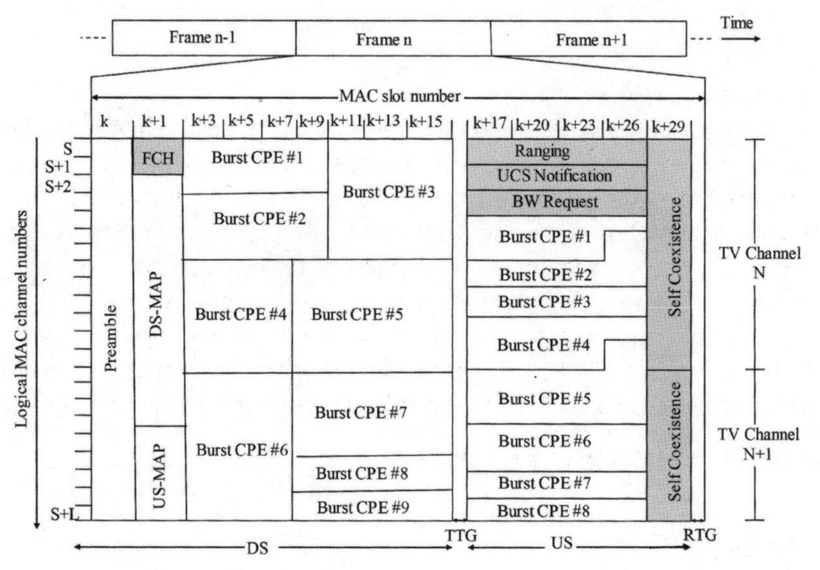

Figure 6.5 Time/frequency structure of a MAC frame.

on a secondary basis where no undue interference is caused to the primary user. It is essential that the system is effectively able to adapt itself around the primary users. A cognitive radio networking is required for achieving this and to provide spectrum sensing and adaptation.

Each station in an 802.22 network is required to perform spectrum sensing. The 802.22 network consists of a BS and a number of client stations, referred to as CPEs. The BS controls when sensing is performed and the results of all spectrum sensing are reported to the BS. The final decision as to the availability of a television channel is made by the base station which can rely on spectrum sensing results, geolocation information and auxiliary information provided by the network manager, to make its final decision about channel availability. Since all sensing results are reported to the base station one can think of spectrum sensing as a signal processing and reporting function. The draft standard does not mandate use of a specific signal processing technique. Instead, it mandates specific sensing performance and a standardized reporting structure.

6.6 IEEE 802.22 Spectrum Measurements

Channel management and spectrum sensing or signal measurements form an important part of the overall 802.22 scheme. The MAC layer within the CPEs carries out many important tasks that enable its smooth and efficient working.

The base station instructs the CPEs to perform periodic measurements in one of two formats;

- *In band spectrum sensing:* The in-band spectrum sensing applies to the channels that are by the BS to communicate with the CPEs. In order for this type of sensing to be undertaken it is necessary for the BS to halt the transmissions on the channel. With a short break of the transmissions, the CPEs can then listen for any other transmissions. When assessing the presence of other signals on the channel, the CPE is required to look for very low level, the levels required and the accuracy being controlled by the BS. The duration of the measurement, which channels, length of measurement time, and probability of false alarm are all under the control of the BS. In order to gain the best overall measurement, the BS may instruct different CPEs to make different measurements. The choice of its modus operands is made by the BS and is calculated by the algorithms it contains. By instructing different CPEs to make different measurements and over different lengths of time, the BS can make up an occupancy map for the overall cell.
- *Out of band spectrum sensing:* Out of band spectrum sensing refers to channels that are not currently being used by the BS for communication with the CPEs. These measurements are made for locating possible alternative channels, should those in use become occupied. It also ensures the presence of a sufficient guard band between the channels in use by the BS and any TVs stations that may be using adjacent channels.

6.7 Turnaround Time Problems

Due to the difference in propagation delay from BS to the CPE, different CPE finish the downlink reception at different points of time. Specifically, a nearby CPE can finish its DL reception long before a faraway CPE does. This also means the readiness of a nearby CPE to start uplink transmission before a faraway CPE does. However the guarantee of reliable reception at the BS requires scheduling of the UL transmissions from different CPE's to enable alignment of OFDMA symbol at the BS. This causes long turnaround time for the TDD method. Turnaround time is the time taken by the BS to send data to CPEs and receive acknowledgements from them indicating the successful receipt of the data by all CPE'S.

6.8 Modified Duplex Technique

The Data transmission done by conventional TDD technique is shown in Figure 6.6. It is clear that the UL transmission of CPE 1 has to wait until CPE 4 completes its UL transmission. It shows the CPE which receives the DL sub frame first sending its UL sub frame at last. This reverse order between DL and UL is the reason for the alignment of OFDMA symbol boundaries at the BS. The nearer CPEs are the most affected by this process of TDD transmission that its turnaround time is long even if they are close to the BS.

The Modified duplex technique involves the following specifications:

- The data transmission to CPEs in the range 1 (nearer to base station) is done by TDD method.
- The data transmission to CPEs in the range 2 (farer to base station) is done by FDD method. As a result, the time taken for data transmission in both the ranges is almost equal. Also the overall turnaround time taken is reduced in comparison with the conventional TDD method.

Figure 6.7 shows the data transmission by modified duplex Technique. Here, TDD technique is used for CPEs 1 and 2 which are nearer to BS and FDD technique for CPEs 3 and 4 which are far from the BS. In FDD, each CPE is allocated a unique channel that the data transmission of one CPE with no needs to wait for other CPEs to complete.

Table 6.2 provides a summary of the simulation parameters. The free space path loss model is considered for simulation. There is also the assumption of

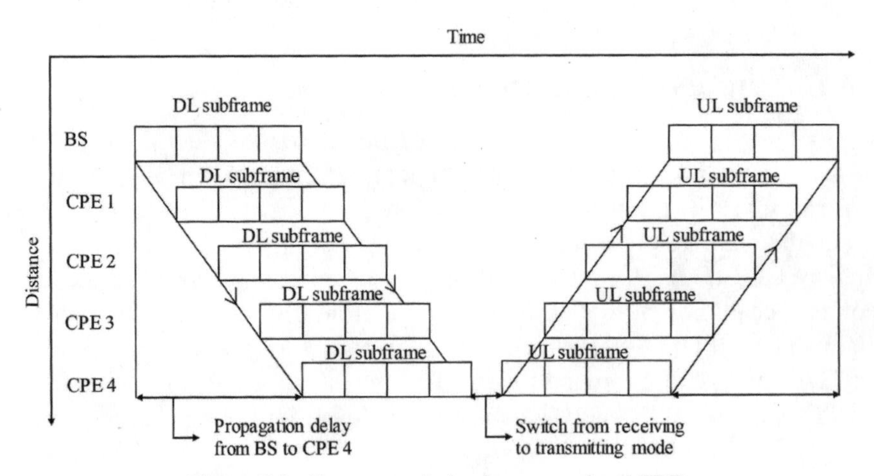

Figure 6.6 Data transmission by conventional TDD.

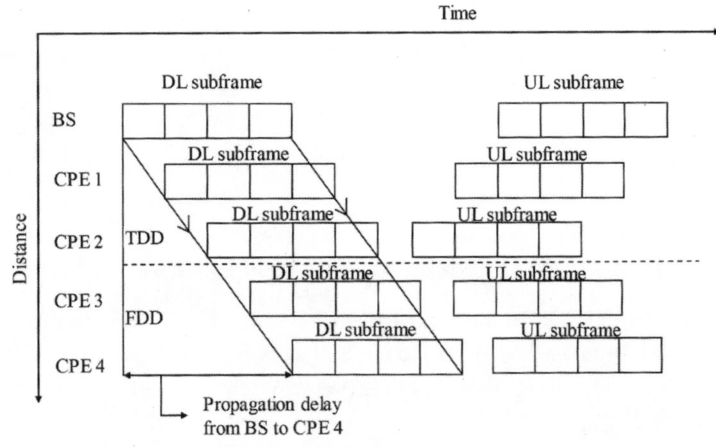

Figure 6.7 Data transmission by modify duplex technique.

Table 6.2 Simulation parameters

Parameter	Specifications
Number of users in the desired cell	40
Number of users in the adjacent cell	80
Number of desired BSs	1
Number of adjacent BSs	2
Multicarrier multiplexing	OFDM
Path loss model	Free space path loss model
Path loss exponent	4
User equipment transmit Power	1 W (30 dBm)
BS transmit power	16 W (42 dBm)
Radius of the cell	17.5 km

40 users per cell considering the fact that this service is provided in the sparsely populated rural areas. The final assumption made considers the interference to have a larger value compared to the noise in the system, neglecting the noise.

6.9 Simulation Results

6.9.1 Representation of the Cells

Figure 6.8 illustrates the representation of cells in the simulation.

Specific coordinates are set for this simulation for the base stations (centre of each cell). Using these coordinates, circles are drawn around the base station representing the TDD and the FDD zones. The users are randomly distributed by randomly generating coordinates in the x and y direction. Those users whose coordinates outside of the circle are given new coordinates till they do not lie inside the circle. Figure 6.8 is a representation of three cells with forty users in each cell. These forty users are randomly distributed in the cell so that some users are in the inner region (TDD) and some are in the outer region (FDD).

6.9.2 Performance of the Modified Duplex System

Figure 6.9, provides a comparison of performance of the TDD with the proposed modified duplex system for the same transmit powers. The performance is obtained by computing the Cumulative distribution function (CDF) for various SIR trails.

The results are obtained for a traditional TDD system and a modified duplex system with an inner radius of 7.5 km. The inference from Figure 6.9 and Table 6.3 is a performance improvement of 9 dB or 39 dBm in the HDD system when compared to the TDD system. Table 6.3 the results obtained in successive iterations. 1000 iterations are performed of which 10 are shown.

The observation from Table 6.3 is that the best SIR performance of 34.0601 dB was obtained for the proposed modified duplex system whereas it was only 25.9875 dB for the TDD system. Also, the worst case performance observed for the modified duplex system had a value of 18.2096 dB which is better when compared to the worst value of the traditional TDD system, which was obtained as 9.4689 dB. This reiterates provides a summary of the earlier result of 9 dB performance improvement.

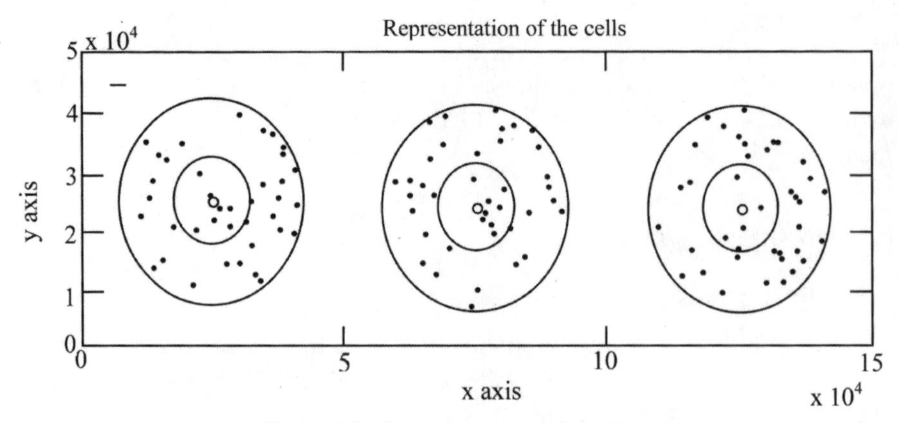

Figure 6.8 Representation of the cells.

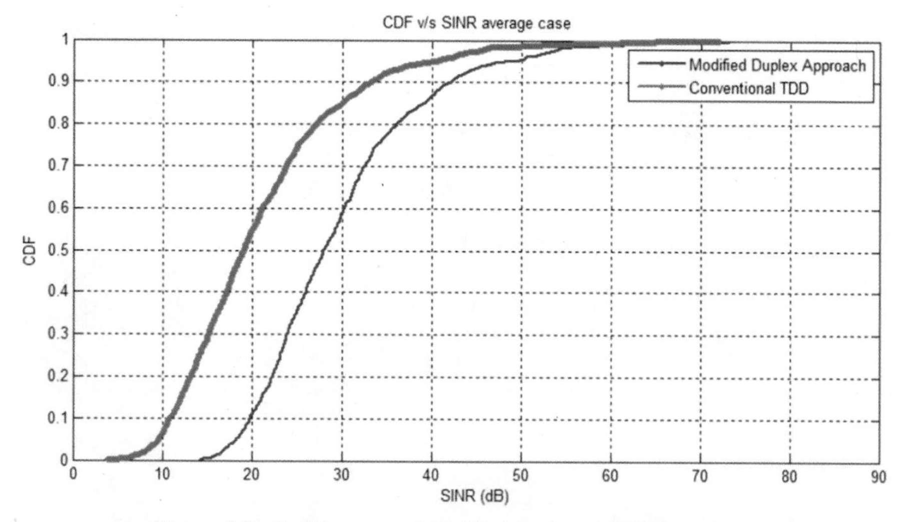

Figure 6.9 Performance of modified duplex and TDD systems.

Table 6.3 SIR values of TDD and modified duplex system

SIR of TDD System (dB)	SIR of Modified Duplex System (dB)
9.4968	18.2096
14.7601	21.7156
23.6467	34.0601
9.4689	17.6122
20.0039	27.7638
25.1401	32.8527
25.9875	35.1888
15.1854	25.7885
13.6543	22.7423
16.7411	25.3445

6.9.3 Variation in the Number of Users

The increase in the number of users triggers increase in the interference in the system. In this article, the number of users has been changed with comparison between the two systems to 20, 40 and 100 and compares the two systems. Figure 6.10 describes the system in reference. The author considers 20 users in each cell for one case and 100 in the other. The number of users was varied for both the TDD as well as modified duplex system. Then the CDF is plotted for both these cases.

The inference from Figure 6.11 is that the performance of the modified duplex system is better than that of the TDD system, showing an 8 dB improvement in the 20 users per cell case and 7 dB in the 100 users per cell case. The performance of the proposed system in the 100 users per cell

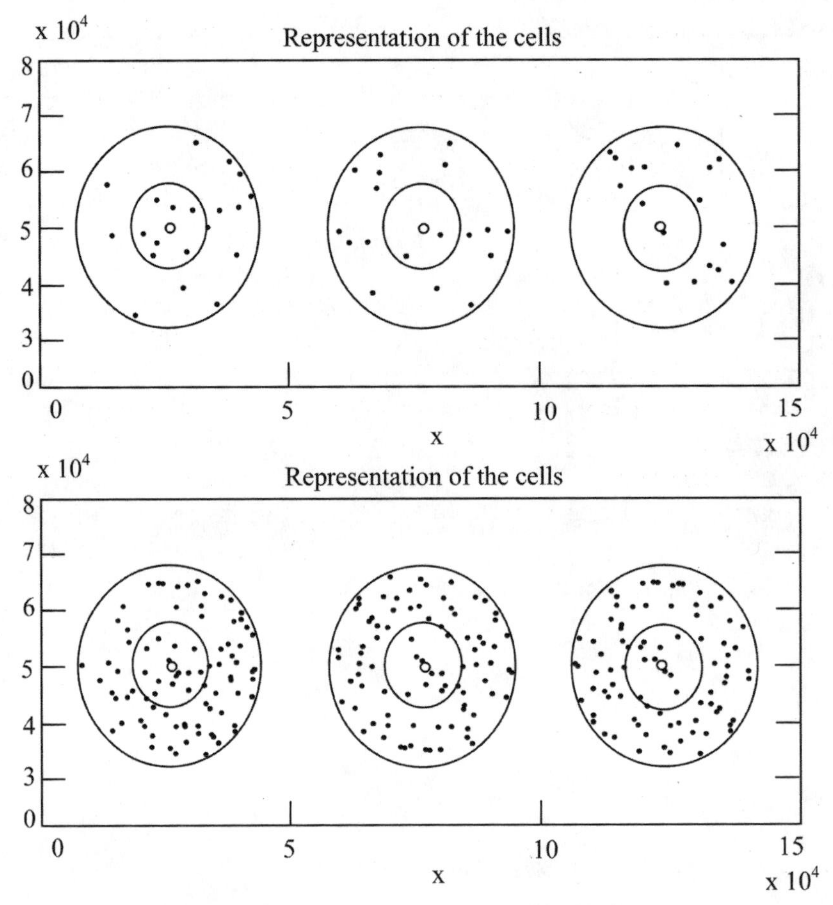

Figure 6.10 Representation of 20 and 100 users.

case is even better than the performance of the TDD system in the 20 users per cell case, leading to the conclusion modified duplex is a better scheme for the WRAN system. The results indicated below show the difference between the performance of TDD and HDD system:

i. 20 users = 9.1000
ii. 100 users = 8.7712

Moreover, increasing the number of users to 100 users from the standard 40 users per cell case resulted in a performance degradation of only 2 dB. Thus, the modified duplex system is not affected very much by the increase in the number of users.

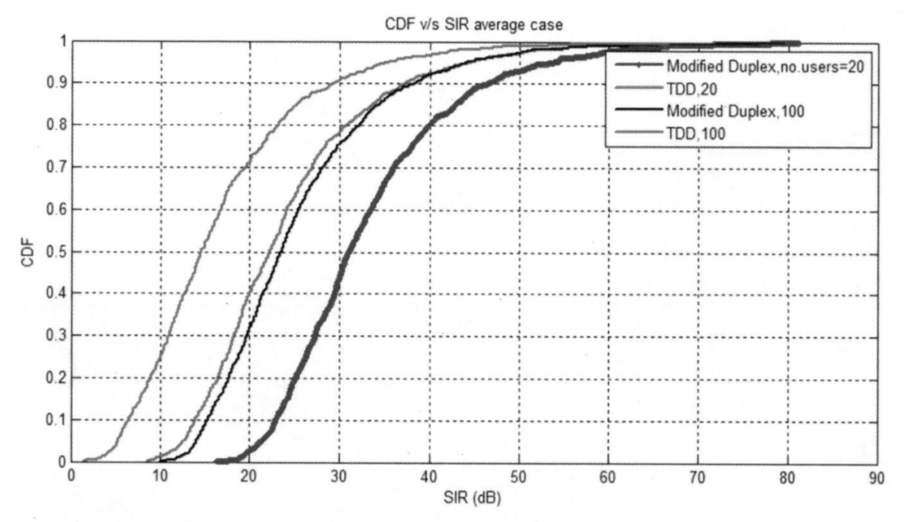

Figure 6.11 CDF v/s SIR for a variable number of users.

6.10 Methodology for Idle Time Calculation

Only TDD (conventional method):

$$Turnaround\ time = \frac{2 * R * 10^3}{(3 * 10^8)} \tag{6.1}$$

where $R =$ range of BS (33 to 100 kms).
TDD and FDD (effective method):

$$\begin{aligned} Turnaround\ time &= TDD\ time\ for\ half\ the\ range + FDD\ time \\ &= \left[\frac{2 * \left(\frac{R}{2}\right) * 10^3}{(3 * 10^8)}\right] + T \end{aligned} \tag{6.2}$$

where $T =$ time taken for sending data to CPEs in range 2 using FDD.

6.11 CTS Interference in IEEE 802.22 Wran Networks

When the frames of a WRAN cell are not synchronized with the adjacent cell, it is severely affected by CTS interference. This is the interference that occurs when two adjacent cells are in different modes during the same time slot, that is, with the simultaneous presence of the cells are in UL and the other adjacent cell is in DL. This interference is present when the system operates in the

TDD mode. Figure 6.12 is a representation of CTS interference. The transmission time slots at the top are for cell 1, whereas those of cell 2 are at the bottom. No CTS interference is encountered. When the time slots of both cells 1 and cell 2 are in UL or DL mode. However, when the slots for both cells are different (depicted by the shaded region), i.e., one cell is in UL mode and the other cell is in DL mode or vice versa, the system suffers from CTS interference.

Therefore, CTS interference arises when the adjacent TDD Base Stations (BSs) have different traffic symmetries or do not synchronize their frames. The illustration of CTS using cell-1 and cell-2 is provided in Figure 6.13. One cell is in UL mode whereas the other is in DL mode. The main interferences include BS-to-BS and MS-to-MS. BS-to-BS interference is a major challenge because it may have a lower path loss exponent compared to BS-MS interference. This poses a serious challenge in TDD-based 4G networks. Location based duplex approach is presented in the next section. This is meant for mitigating this interference.

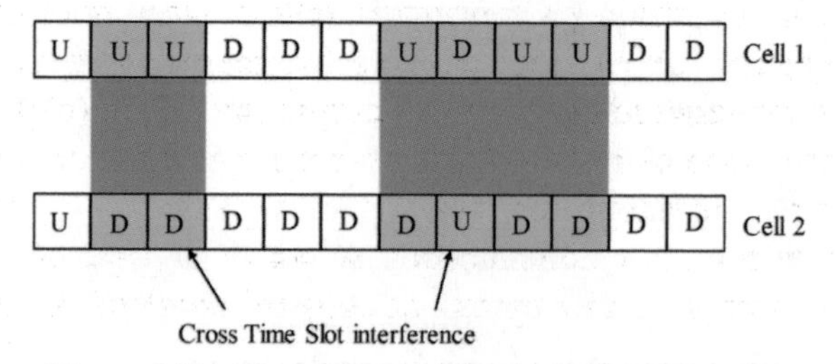

Figure 6.12 Illustration of cross time slot interference.

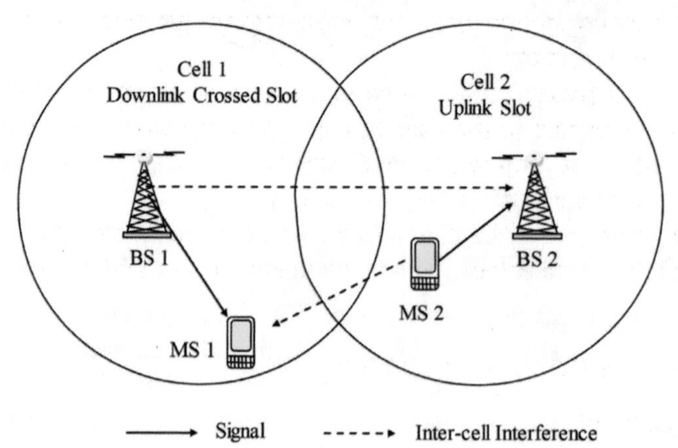

Figure 6.13 Cross time-slot interference problem.

6.12 CTS Interfrence Mitigation in IEEE 802.22 Wran Networks

6.12.1 System Model and Its Elements

The different elements of the systems are

i. BS: It is the base station in our system model. The author have kept one BS as the desired Base Station (the center one among the three), with the other two BSs as interference causing.
ii. aMS: Adjacent Mobile Station (aMS) stands for the mobile stations in the cells adjacent to the desired cell. The adjacent mobile stations are responsible for the interference to the desired mobile stations and the desired base station.
iii. dMS: Desired Mobile Station (dMS) stands for the mobile stations in the desired cell. The authors intend reducing the interference in the desired cell with desired base station and mobile station.

The system is considered as a cluster of cells for studying the interference characteristics in cellular communication where as in IEEE 802.22 system, tier by tier architecture is followed .This shown in Figure 6.14a. Consider a TV transmitter surrounded by three WRAN BS cells. The distance from the base station of TV transmitter to the boundary of that cell region (d_{DTV}) is added with the safe distance (d_{safe}) which further separates the boundary of different cell regions from the inside TV transmitter region. Now, in order to reduce the interference between the three cells outside the TV transmitter region, it is provided with a guard distance (d_{guard}) between them. Consider the middle cell of the WRAN system as the desired cell and the other two surrounding cells as the interference causing cells. Hence, only the interference caused to the WRAN BS2 cell is taken into account and the interference caused to the WRAN BS1 and WRAN BS3 can be neglected.

WRAN BSs can be employed beyond the distance $(d_{DTV} + d_{safe})$ where (d_{DTV}) is the DTV service range and (d_{safe}) is the protection range (d_{DTV}) and (d_{safe}) are selected according to FCC rules reported in [Gerald Chouinard 2005]). Furthermore, d_{WRAN} denotes the WRAN BS service coverage area and d_{guard} denotes the distance between adjacent WRAN BSs.

The interference caused in cell 2 (the desired cell) is due to its two adjacent cells, cell 1 and cell 3. When the TDD frames of the cells are not synchronized with each other, MS-to-MS and BS-to-BS interference is caused between the cells. Such interference is termed as CTS interference. Each cell is divided into two regions for reducing such CTS interference inner part operating by

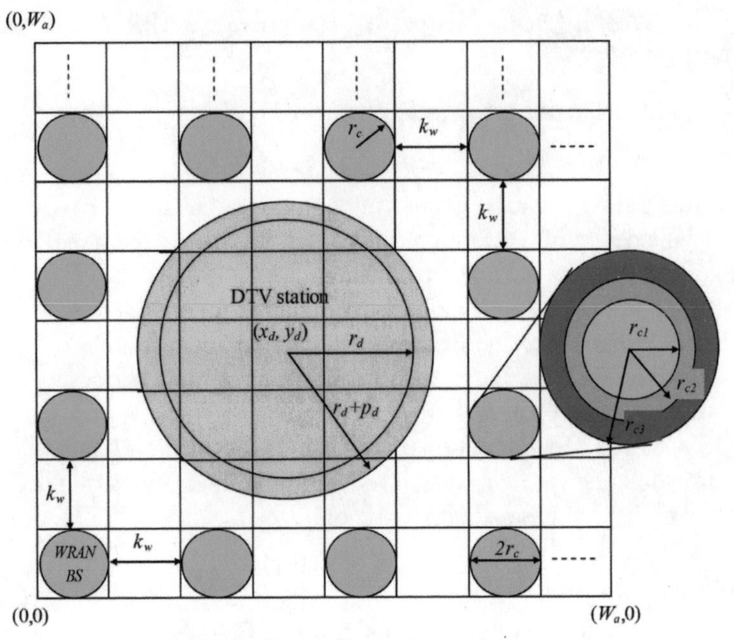

Figure 6.14(a) Tier by tier architecture.

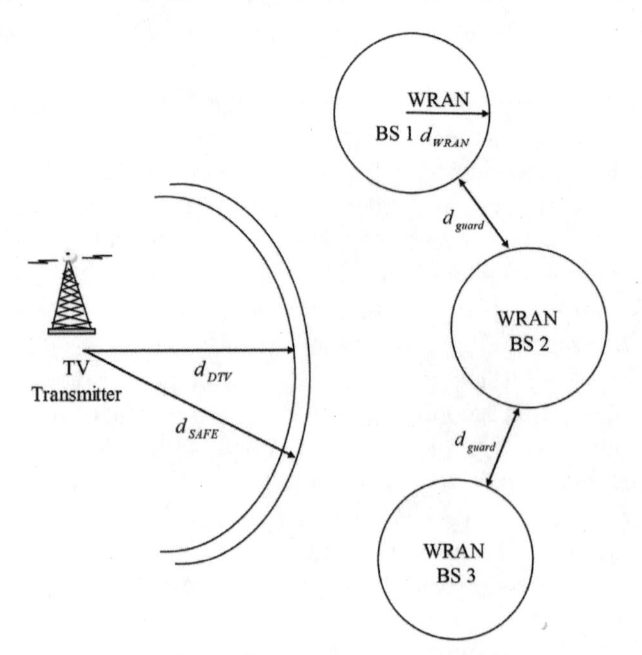

Figure 6.14(b) WRAN system model.

TDD in UL as well as in DL and the outer part operating by TDD in DL and FDD in UL. In a conventional FDD system, a pair of spectrum bands is used for UL and DL, but the LBD scheme utilizes FDD in DL mode only. Thus, simultaneous operation of TDD and FDD is achievable in a single cell without any need of additional spectrum.

Alternatively the inner part of the cell is used for operation by FDD in UL and TDD in DL and the outer part operating by TDD in UL as well as in DL. But this type poses a large turnaround time problem for outer part of the user.

6.13 Interference Scenarios

It is assumed that the system utilizes a frequency currently unused by the DTV in the particular area. For the sake of simplicity, the authors consider only one adjacent cell. Two particular cases: i) the desired cell in UL ii) the desired cell in DL.

6.13.1 Desired Cell in Uplink

Figure 6.15 shows the scenario, where the desired cell is in the UL mode and the adjacent cell is in the DL mode. Therefore BS of the adjacent cell is doing the transmission while the desired cell is receiving information, resulting in

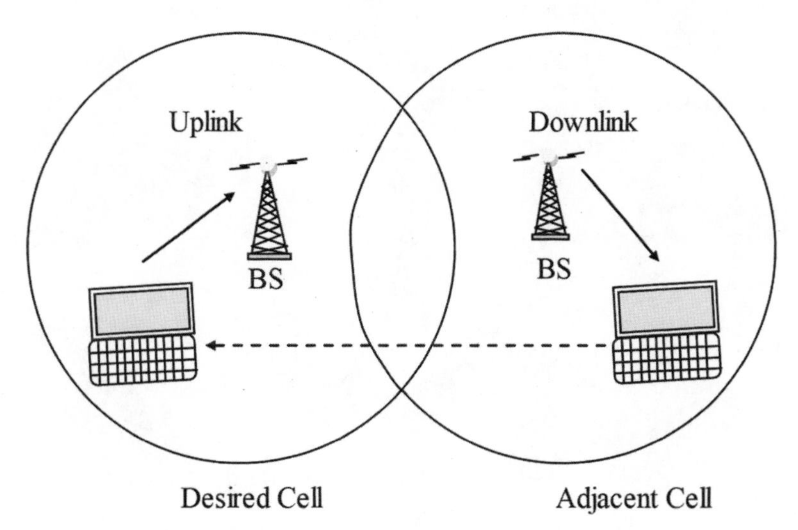

Figure 6.15 Interference when desired cell is in UL.

BS-to-BS interference (depicted by dashed lines). Such interference usually affects the system more than the MS-to-BS interference because a line-of-sight (LOS) path may exist between the elevated BSs and the path loss exponent that exists between elevated BSs can be smaller than that between BS and MS.

6.13.2 Desired Cell in Downlink

Figure 6.16 depicts the interference scenario when the desired cell has its TDD slot in DL and the adjacent cell has is TDD slot in UL. The desired MS, receiving information from its BS, is susceptible to interference from the mobile station in the adjacent cell (depicted by dashed lines). This interference is tough to mitigate compared to the former case as the location of the MSs might keep changing.

The CTS interference poses a problem to the system that uses TDD, reducing the throughput of the system. Different approaches were considered for reduction in the interference in such a system. FDD is another alternative to TDD since FDD has the benefit of requiring no guard time between UL and DL. However, the problem with FDD is that it does not support asymmetrical data services. If the cross slots interference so strong, it may affect adjacent channels noticeably. Hence FDD cannot be implemented on a large scale due to the above problems.

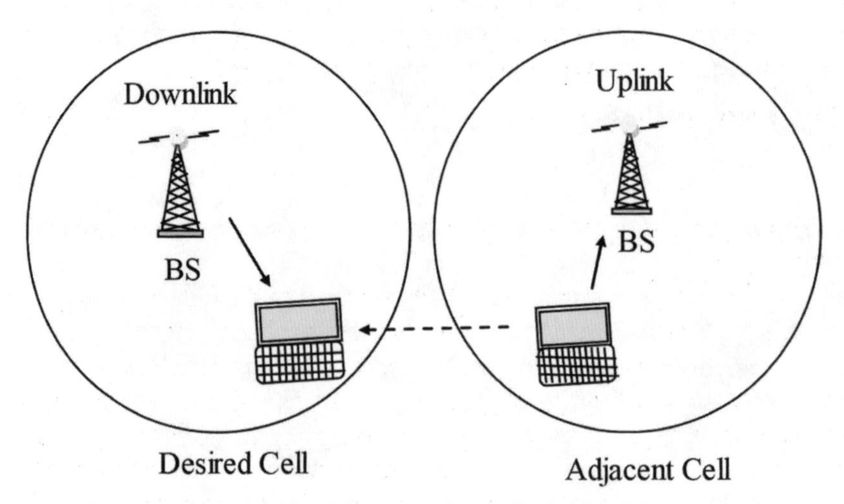

Figure 6.16 Illustration when desired cell in DL.

6.14 Location Based Duplex Scheme

Figure 6.17 gives a description of the LBD system. In LBD scheme, entire cell region is divided into two parts, namely, the inner part and the outer part. The inner part of the cell operates by TDD in UL and DL, while outer part operates by FDD in UL and TDD in DL. The important advantage of the proposed scheme is its robustness against large turnaround time and CTS interference that is inherent in the TDD system, which is caused by the asynchronous downlink/uplink switching boundaries among all neighbor cells.

Table 6.4 describes the various parameters of system based on the BS EIRP (for 100 [W] and 4 [W] separately).

6.14.1 Mathematical Model

Case 1: Desired cell in DL/Adjacent in UL (worst case)
Let us assume the MS is located at the edge of cell 2 as shown in Figure 6.18. Cells 1 and 3 operate in the UL mode while Cell 2 operates in the DL mode.

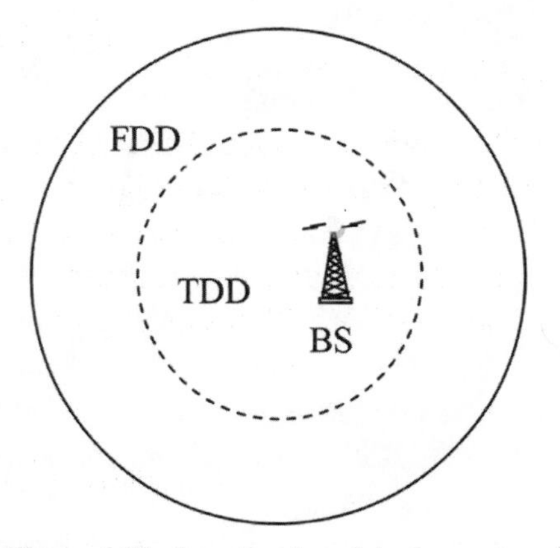

Figure 6.17 Location based duplex system.

Table 6.4 System parameters with respect to BS EIRP

Base Station EIRP	100 [W]	4 [W]
	20.0 [dBW]	6.0 [dBW]
Antenna height above average terrain (HAAT)	75.0 [m]	75.0 [m]
Required Field Strength	28.8 [dB(uV/m)]	28.8 [dB(uV/m)]
Path loss needed beyond 1 m	126.0 [dB]	112.0 [dB]
Radius of the cell	30.3 [km]	16.7 [km]

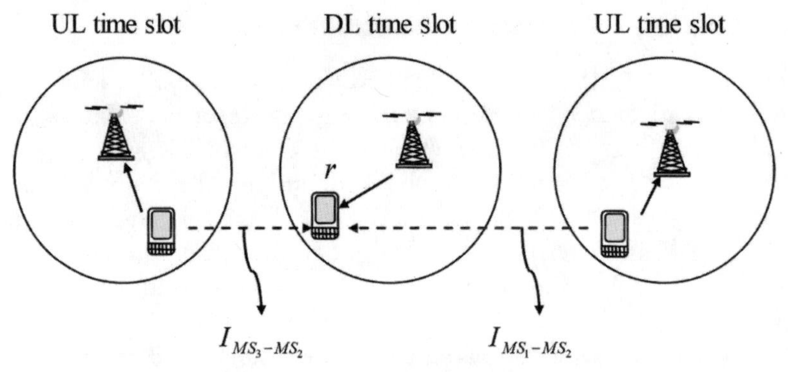

Figure 6.18 Worst case scenario in TDD based WRAN.

The signal power received by MSs in cell 2 from the BS in cell 2 is given by

$$S^{BS \to MS} = \alpha \mu_{BS} P_{BS}(r)^{-\gamma} \tag{6.3}$$

where γ is the path loss exponent, α is a constant, P_{BS} is the power transmitted by the BS and μ_{BS} is cell loading factor of the BS, which is defined as the ratio of the number of used subcarriers to the total number of subcarriers in the OFDMA system.

In the traditional system, cells 1 and 3 (operating in TDD-UL) cause interference in cell 2 (operating in the TDD-DL). The interference caused to any MS in cell 2 by the MSs in the adjacent cells is given by

$$I_{DL}^{MS \to MS} = \sum_{i} \alpha \mu_{MS} P_{MS}(r_i)^{-\gamma} \tag{6.4}$$

where 'i' is the total number of MSs in the adjacent cells and r_i is the distance between the MS in cell 2 and the ith MS in the adjacent cell. The overall SIR of the MS with distance from the BS is r for the conventional system is given as

$$SIR_{conv} = \frac{S_{BS \to MS}}{I_{MS_1^i \to MS_2} + I_{MS_1^0 \to MS_2} + I_{MS_3^i \to MS_2} + I_{MS_1^0 \to MS_2}} \tag{6.5}$$

where $I_{MS_1^i \to MS_2}$ is the interference from MS located inside the cell 1-to-cell 2 (target cell), $I_{MS_1^0 \to MS_2}$ is the interference from MS located outside of cell 1-to-cell 2, $I_{MS_3^i \to MS_2}$ is the interference from MS located inside the cell 3-to-cell 2 and $I_{MS_3^0 \to MS_2}$ is the interference from MS located outside of cell 3-to-cell 2.

Equation (6.5) shows the SIR that affects from the interference by due to the MSs located in the adjacent cell. When many MSs are located at the edge of

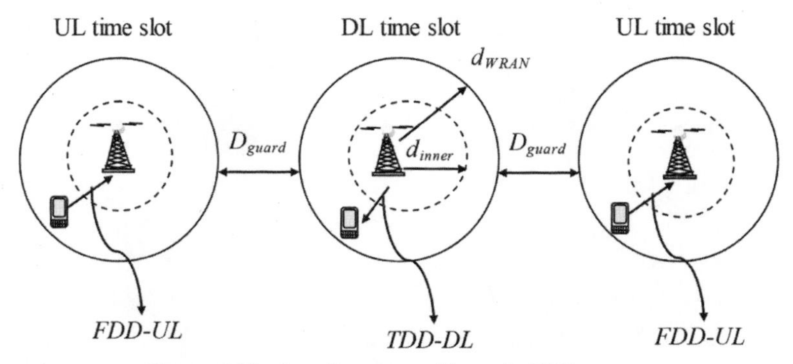

Figure 6.19 Interference avoidance in LBD system.

the adjacent cell, the performance degrades. However, in the LBD system, for UL transmission, MSs located in the outer region of the cells operate by FDD mode and hence the MS-to-MS CTS interference is considerably reduced. This is illustrated in Figure 6.19. The overall SIR of the MS in location based scheme is given as

$$SIR_{LBD} = \frac{S_{BS \to MS}}{I_{MS_1^i \to MS_2} + I_{MS_3^i \to MS_2}} \tag{6.6}$$

Equation (6.6) shows the performance of the LBD scheme affected by MS located in the inner region of an adjacent cell. However, the effect of interference from MSs located in the inner regions of the cell to the desired MS is neglected as the transmission power of MS located nearer the BS is always low. Comparing Equations (6.3) and (6.4), the LBD system produces better SIR performance.

6.15 Performance Analysis

Table 6.5 summarizes the simulation parameters while free space path loss model is considered for simulation. In the simulation, 40 users per cell are taken considering the fact that this service is provided in the sparsely populated rural areas. In addition, it is consider that interference dominates compared to the noise in the system.

Figure 6.20 shows three cells with forty users in each. These forty users are randomly distributed in the cell with some users in the inner region (TDD) and the other in the outer region (FDD). Figure 6.21 shows a comparison of the interference performance of the conventional TDD with the proposed LBD

Table 6.5 Simulation parameters

Parameter	Specifications
Number of users in the desired cell	40
Number of users in the adjacent cell	80
Number of desired BSs	1
Number of adjacent BSs	2
Multicarrier multiplexing	OFDM
Path loss model	Free space path loss model
User equipment transmit Power	1 W
BS transmit power	16 W
Radius of the cell	17.5 km

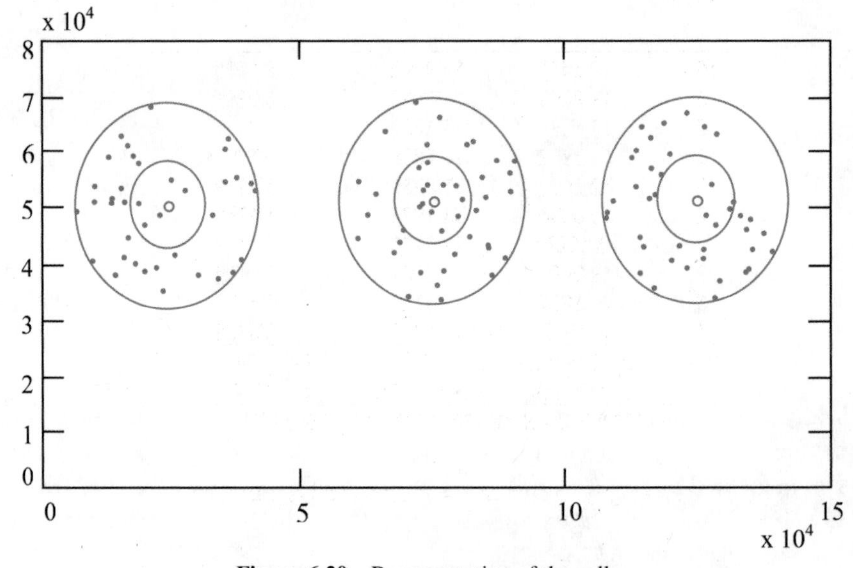

Figure 6.20 Representation of the cells.

system for the same transmit powers for 1000 iterations. The performance is obtained by computing the cumulative distribution function (CDF) for the various SIR trails.

The results obtained are for a conventional TDD system and a LBD system for an inner cell radius of 7.5 km. The interference from Figure 6.21 and Table 6.6 is that there is a performance improvement of 9 dB or 39 dBm in the LBD system when compared to the TDD system. Table 6.6 summarizes the simulation results obtained during different iterations. Table 6.4 leads to the observation that the best SIR performance of 34.0601 dB was obtained for the proposed LBD system whereas it was only 25.9875 dB for the

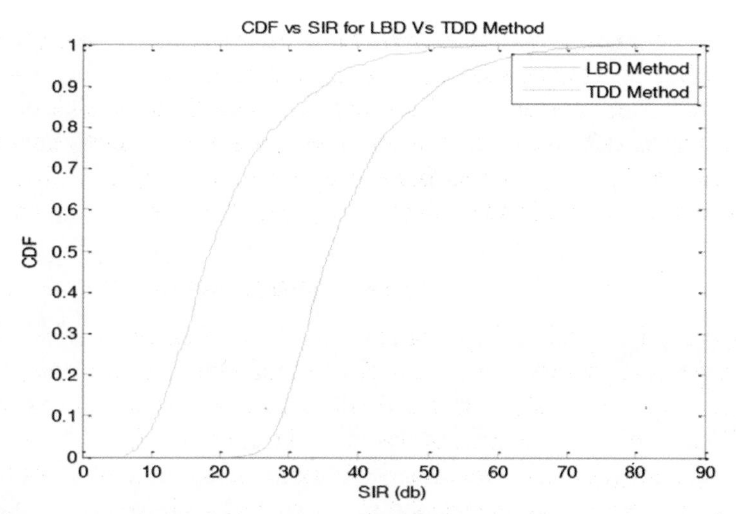

Figure 6.21 Performance of LBD and TDD systems.

Table 6.6 SIR values of TDD and LBD system

S.No.	SIR of TDD System (dB)	SIR of LBD System (dB)	Performance Improvement (dB)
1	9.4968	18.2096	8.7128
2	14.7601	21.7156	6.9555
3	23.6467	34.0601	10.4134
4	9.4689	17.6122	8.1433
5	20.0039	27.7638	7.7599
6	25.1401	32.8527	7.7126
7	25.9875	35.1888	9.2013
8	15.1854	25.7885	10.6031
9	13.6543	22.7423	9.088
10	16.7411	25.3445	8.6034

TDD system. The worst case performance observed for the LBD system had a value of 18.2096 dB which is better when compared to the worst value of the traditional TDD system, which was obtained as 9.4689 dB. This reiterates the results reported by the authors that there is a 9 dB performance improvement.

6.15.1 CDF for Varying Inner Cell Radii

The inner cell radius is one of the most important parameters in LBD system influencing the ratio of the area using the TDD scheme to that using the FDD scheme. The ratio of the inner cell radius to the entire cell radius varies from

0.25 to 1. One Thousand SIR trials are performed for each case and a CDF is plotted for the same. From Figure 6.22, it can be inferred that the best performance is obtained for an inner cell radius of 4.375 km. For the case where the inner cell radius is 17.5 km, the performance is worst and is the same as that of the traditional TDD system.

The performance improves as the inner cell radius decreases. In spite of that, designing a LBD system with a very small radius is not advised as most of the regions will be outside the TDD region, making it an FDD system altogether. Hence the optimal radius should be between 8.75 km and 11 km to ensure that enough number of users lie in the inner region.

6.15.2 Variation of the Inter-Cell Distance

In this part a study is made of the effect of guard distance (d_{guard}) between two cells, called inter-cell distance. A noise limited system is achieved with increase in the inter cell distance. But alternatively a large inner cell distance reduces the throughput of the system. Since smaller number of

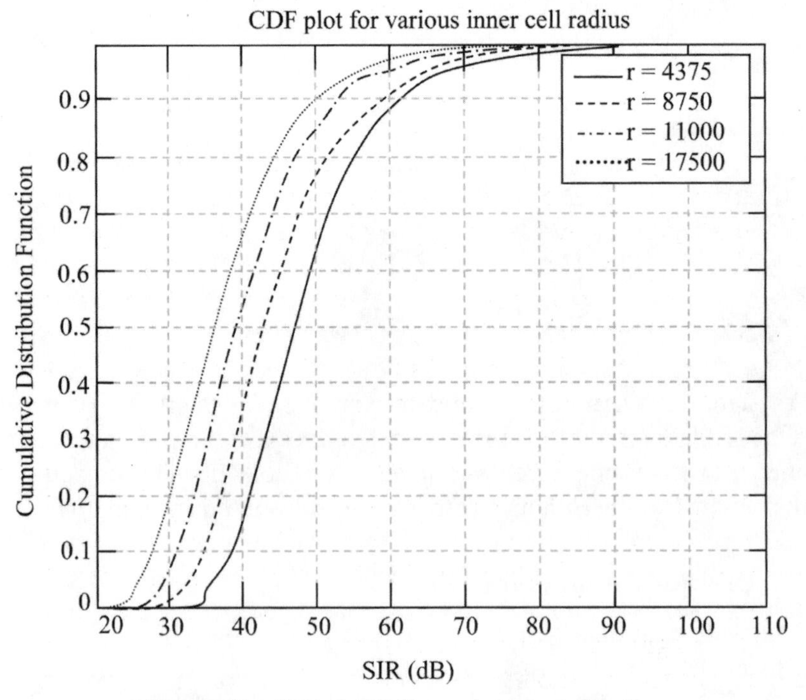

Figure 6.22 CDF v/s SIR for varying inner cell radius.

frequency reuse is possible. In the conventional TDD system, a distance of $d_{guard} = r_c$ between the cells is maintained (where r_c is the radius of the cell in consideration). Such a large inter-cell distance reduces the throughput of the system when LBD is applied to the system; the performance is similar to the TDD system when the distance is reduced to $r_c/2$. Figure 6.23 shows reduction in the inter cell distance to half without affecting the performance.

6.15.3 Variation in the Number of Users

The increase in the number of users causes the interference in the system. Here, the change in the number of users to 20 and 100 is considered and a comparison of the performance of the two systems is made. The number of users was varied for both the TDD as well as LBD system. Then the CDF is plotted for both these cases. From the interference from Figure 6.24 is that the performance of the LBD system is better than that of the TDD system, showing an 8 dB improvement in the 20 users per cell case and 7 dB in the 100 users per cell case. Furthermore, the performance of the LBD system in the 100 users per cell case is even better than the performance of the TDD system in the 20 users per cell case, this proves that LBD is a better scheme for the WRAN system. Moreover, increasing the number of users to 100 from the standard 40 in per cell case results in a performance degradation of just 2 dB.

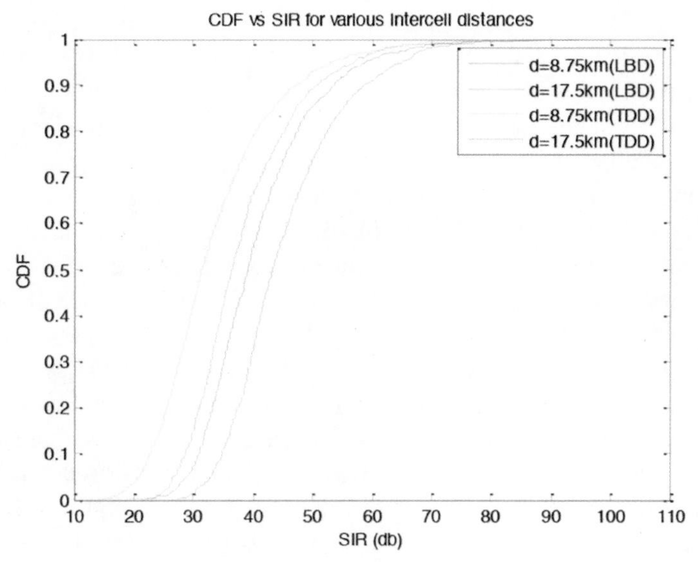

Figure 6.23 Inter cell distance variation.

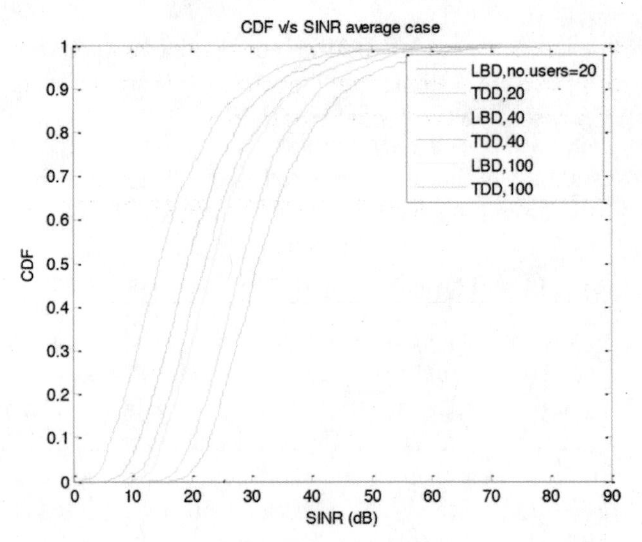

Figure 6.24 CDF v/s SIR for a variable number of users.

Thus, the LBD system is not affected substantially by the increase in the number of users.

6.15.4 Single User in the Desired Cell

A single user at the edge of the cell is considered for analysis of a scenario of the worst case. As the user is located at the edge of the cell, it receives maximum interference from the users in the adjacent cells. A performance improvement of 9–10 dB is obtained in the system. Since the user is at the edge, the SIR is of a small value when compared to the average cases. The improvement of SIR is similar to the earlier result of the average case. Figure 6.25 provides the CDF comparison of the TDD and LBD systems for a single user.

Figure 6.26, indicates the performance of a single user at the cell edge. A comparison of the performance of various inner cell radius is made. It has been observed that optimal radius should be 8.75–11 km for better performance.

Figure 6.27 provides a comparison of the proposed LBD system with existing method to solve CTS interference. Literature provides details of various methods for solving CTS, of which the virtual cell approach is considered good. Hence the proposed LBD approach is compared with virtual cell concept in IEEE 802.22 system. The observation from Figure 6.27 is that performance improvement of 9 dB for LBD method compare to virtual cell method has been achieved.

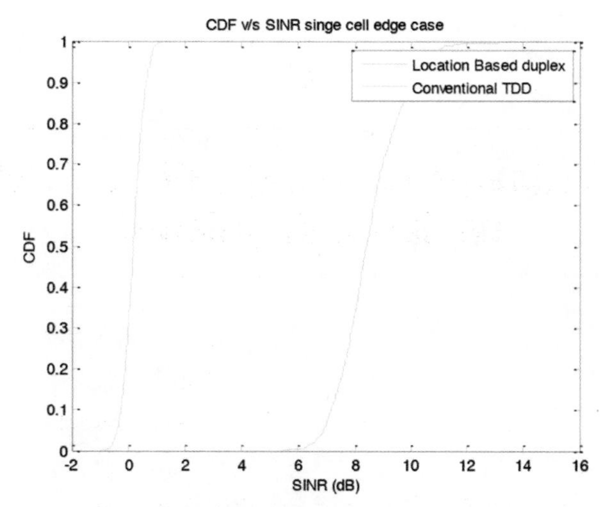

Figure 6.25 TDD v/s LBD for a single user.

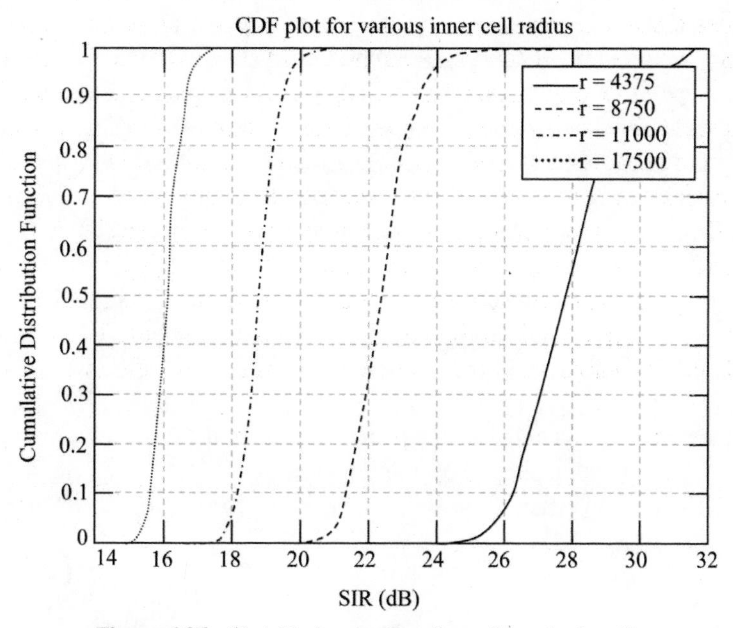

Figure 6.26 Variable inner cell radiuses for a single cell.

Case 2: Desired cell in UL/Adjacent cell in DL

Figure 6.28 illustrates the system model for case 2, in this case Cell 2 (desired cell) operates in UL mode and the adjacent cells operate in DL mode. In this

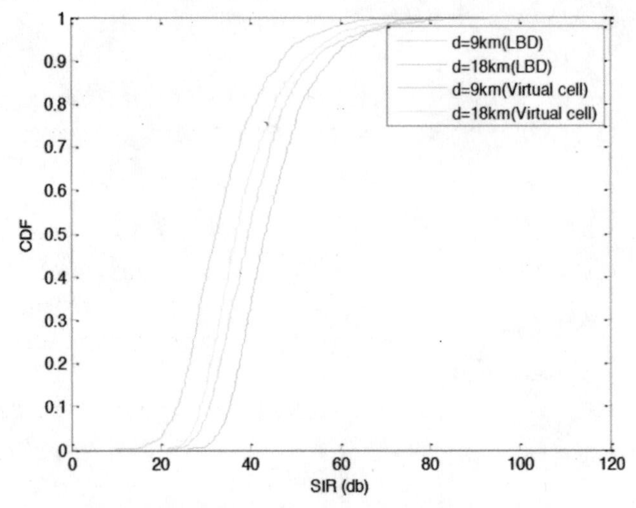

Figure 6.27 Performance comparison of LBD Vs virtual cell concept.

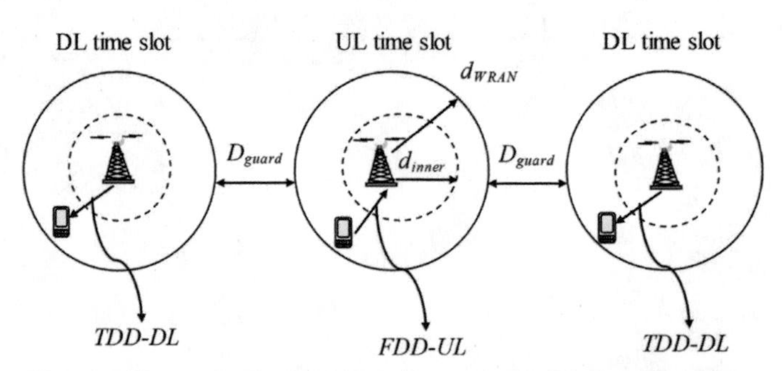

Figure 6.28 Desired cell in UL/adjacent cell in DL in LBD system.

case, the desired user in cell 2 is free from CTS, since the desired users operate by FDD in UL and adjacent cell user are operated by TDD-DL mode. As these two signals have no correlation with each other, better SIR performance is achieved using the LBD system.

7

MATLAB Programs for Spectrum Sensing Technqiues

7.1 Energy Detection

AIM:

To detect the unknown PU (Primary user) signal based on energy of the received signal. The Energy detector is optimal to detect the unknown signal if the noise power is known. In the energy detection, CR users sense the presence/absence of the PUs based on the energy of the received signals. As above, the measured signal r (t) is squared and integrated over the observation interval T. Finally, the output of the integrator's compared with a threshold k to decide if a PU is present. Since the energy detection depends only on the SNR of the received signal, its performance is susceptible to uncertainty in noise power. If the noise power is uncertain, the energy detector will not be able to detect the signal reliably as the SNR is less than a certain threshold, called an SNR wall. In addition, the energy detector can only determine the presence of the signal but cannot differentiate signal types. Thereby, the energy detector often results in false detection triggered by the unintended CR signals.

MATLAB CODE:

i) Code for performance of detector (Pd) under various values of probability of false alarm (pfa) for SNR = 4 db

```
clc;
clear all;
close all;
m=10;
N=2*m;
pf=[0.0001 0.001 0.01 0.1 1];
snr_avgdB=4;
```

```
snr_avg=power(10,snr_avgdB/10);
for i=1:length(pf)
    over_num=0;
    th(i)=gammaincinv(1-pf(i),2*m);
    for kk=1:1000
        t=1:N;
        x=sin(pi*t);
        noise=randn(1,N);
        pn=mean(noise.^2);
        amp=sqrt(noise.^2*snr_avg);
        x=amp.*x./abs(x);
SNRdB_Sample=10*log10(x.^2./(noise.^2));
        signal=x(1:N);
        ps=mean(abs(signal).^2);
        Rev_sig=signal+noise;
        accum_power(i)=sum(abs(Rev_sig))/pn;
        if accum_power(i)>th(i)
            over_num=over_num+1;
        end
    end
    pd_sim(i)=over_num/kk;
end
figure
s=plot(pf,pd_sim,'b-*');
set(s,'linewidth',2);
title('Nonfluctuating Coherent ROC');
grid on;
xlabel('probability of falsed alarm(pfa)');
ylabel('probability of detection(pd)');
legend(['SNR=' num2str(snr_avgdB) 'dB']);
```

(ii) Code for performance of detector under various values of signal to noise ratio (SNR)

```
clc;
clear all;
close all;
m = 10;
```

```
N = 2*m;
base=0.01:0.02:1;
pd_sim = [];
base = 0.1;
pf=base.^2;
for snr_avgdB=0:2:30
    snr_avg=power(10,snr_avgdB/10);
    for i=1:length(pf)
        over_num=0;
        th(i)=gammaincinv(1-pf(i),2*m);
        for kk=1:1000
            t=1:N;
            x=sin(pi*t);
            noise=randn(1,N);
            pn=mean(noise.^2);
            amp=sqrt(noise.^2*snr_avg);
          x=amp.*x./abs(x);
SNRdB_Sample=10*log10(x.^2./(noise.^2));
            signal=x(1:N);
            ps=mean(abs(signal).^2);
            Rev_sig=signal+noise;
            accum_power(i)=sum(abs(Rev_sig))/pn;
            if accum_power(i)>th(i)
                over_num=over_num+1;
            end
        end
        pd_sim = [pd_sim over_num/kk];
    end
end
figure;
SNR = 0:2:30;
s=plot(SNR,pd_sim,'-*');
set(s,'linewidth',2);
title('Energy detection method');
grid on;
xlabel('Signal-to-noise ratio (dB)');
ylabel('probability of detection(pd)');
```

RESULTS:

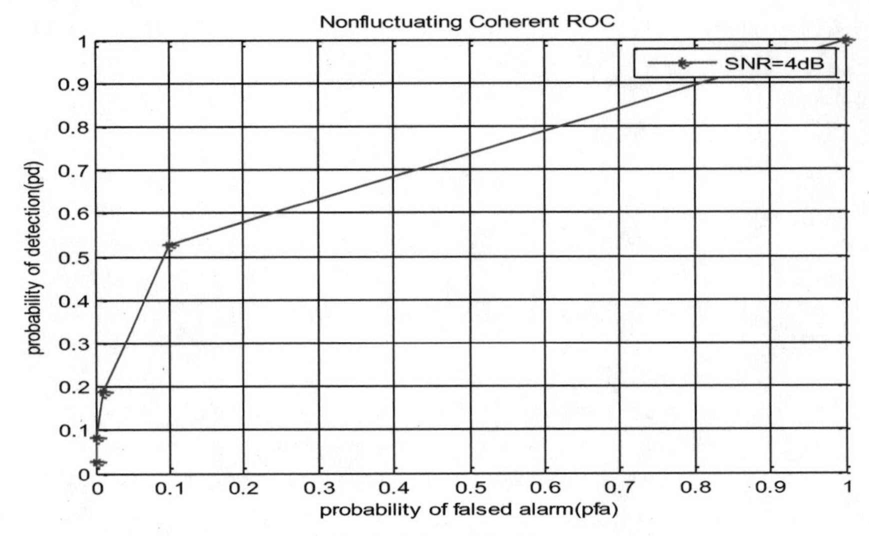

Figure 7.1 Pd Vs Pfa.

INFERENCE:

The observation from the Figure 7.1 is the presence of a tradeoff between Pd and Pfa values. For Pfa values from 0.09 to 0.8 the detection probability is optimum. After that the detection probability approaches 1 for SNR = 4 db.

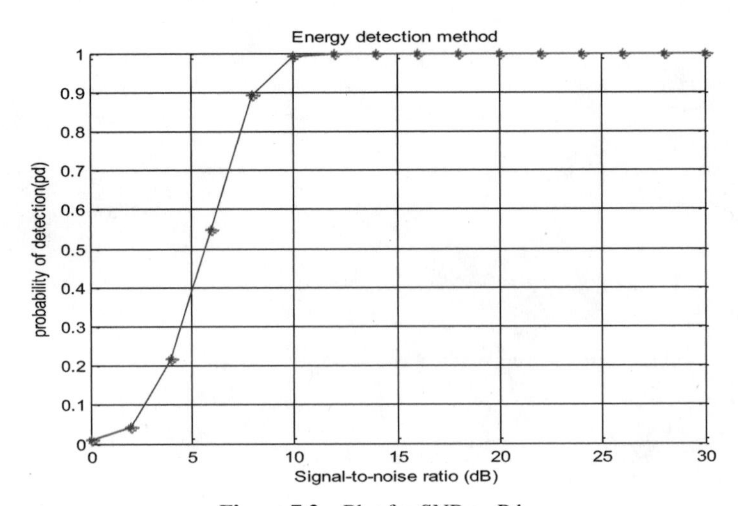

Figure 7.2 Plot for SNR vs Pd.

INFERENCE:

The observation from the Figure 7.2 is the presence of a linear increase in the probability of detection for increases in SNR value. For low values of SNR the detection probability is almost 0. Above 8 db the detection probability approaches 1.

7.2 Matched Filter Detection

AIM:

To detect the unknown PU (Primary user) signal based on Prior knowledge (bandwidth, operating frequency, modulation type etc) and to maximize the SNR. When the user has information on the primary user signal, the optimal detector in stationary Gaussian noise happens to be the matched filter as it maximizes the reclosed signal-to-noise ratio (SNR). The palpable advantage of matched filter lies in its requirement of smaller time for achieving high processing gain for the reason of coherence and of a prior knowledge of the basic user signal like the modulation type and order, the shape of the pulse and the packet format. The performance of the matched filter will be poor if this information has no accuracy. Most wireless networks system have pilot, preambles, synchronization word or spreading codes and hence can be used for coherent detection.

MATLAB CODE:

(i) Code for performance of detector (Pd) under various values of probability of false alarm (pfa) for SNR = 2 db

```
clc;
clear all;
close all;
%%
% fix the random number generator
rstream = RandStream.create('mt19937ar','seed',2009);
Ntrial = 1e5;                % number of Monte-Carlo trials
spower = 1;                  % signal power is 1
pd_graph = [];
for snrdb = 2:2:10                   % SNR in dB
    snr = db2pow(snrdb);          % SNR in linear scale
    npower = spower/snr;              % noise power
    namp = sqrt(npower/2);        % noise amplitude in each
                                    channel
```

```
    s = ones(1,Ntrial);          % signal
    n = namp*(randn(rstream,1,Ntrial)+1i*randn
        (rstream,1,Ntrial));  % noise
    x = s + n;
    mf = 1;
    y = mf'*x;
    z = real(y);
    Pfa = 1e-3;
    snrthreshold = npwgnthresh(Pfa, 1,'coherent');
    mfgain = mf'*mf;
    % To match the equation in the text above
    % npower - N
    % mfgain - M
    % snrthreshold - SNR
    threshold = sqrt(npower*mfgain*snrthreshold);
    Pd = sum(z>threshold)/Ntrial
    x = n;
    y = mf'*x;
    z = real(y);
    Pfa = sum(z>threshold)/Ntrial
    [Y, X] = rocsnr(snrdb,'SignalType',
    'Nonfluctuatingcoherent','MinPfa',1e-4);
    pd_graph = [pd_graph; Y'];
end
%%
figure;
%s=semilogx(X,pd_graph(1,:),'r',X,pd_graph(2,:),
'g',X,pd_graph(3,:),'b',X,pd_graph(4,:),
'c',X,pd_graph(5,:),'m');
s=semilogx(X,pd_graph(3,:),'b');
set(s,'linewidth',2);
title('Nonfluctuating Coherent ROC - Matched Filter');
grid on;
xlabel('probability of falsed alarm(pfa)');
ylabel('probability of detection(pd)');
snr_avgdB = 2:2:10;
legend(['SNR=' num2str(snr_avgdB(1)) 'dB'],['SNR='
num2str(snr_avgdB(2)) 'dB'],['SNR=' num2str(snr_avgdB(3))
'dB'],['SNR=' num2str(snr_avgdB(4)) 'dB'],['SNR='
num2str(snr_avgdB(5)) 'dB']
```

(ii) Code for performance of detector under various values of signal to noise ratio (SNR)

```
clc;
clear all;
close all;
%%
% fix the random number generator
rstream = RandStream.create('mt19937ar','seed',2009);
spower = 1;                  % signal power is 1
Ntrial = 1e5;                % number of Monte-Carlo trials
s = ones(1,Ntrial);          % signal
mf = 1;
Pfa = 0.1;
Pd = [];
for snrdb = 0:2:30                        % SNR in dB
    snr = db2pow(snrdb);          % SNR in linear scale
    npower = spower/snr;               % noise power
    namp = sqrt(npower/2);        % noise amplitude in
                                    each channel
    n = namp*(randn(rstream,1,Ntrial)+1i*randn
    (rstream,1,Ntrial));  % noise
    x = s + n;
    y = mf'*x;  % apply the matched filter
    z = real(y);
    snrthreshold = npwgnthresh(Pfa, 1,'coherent');
    mfgain = mf'*mf;
    % To match the equation in the text above
    % npower - N
    % mfgain - M
    % snrthreshold - SNR
    threshold = sqrt(npower*mfgain*snrthreshold);
    Pd = [Pd sum(z>threshold)/Ntrial];
end
figure;
SNR = 0:2:30;
s=plot(SNR,Pd,'-*');
set(s,'linewidth',2);
title('Matched Filter');
```

```
grid on;
xlabel('Signal-to-noise ratio (dB)');
ylabel('probability of detection(pd)');
```

RESULTS:

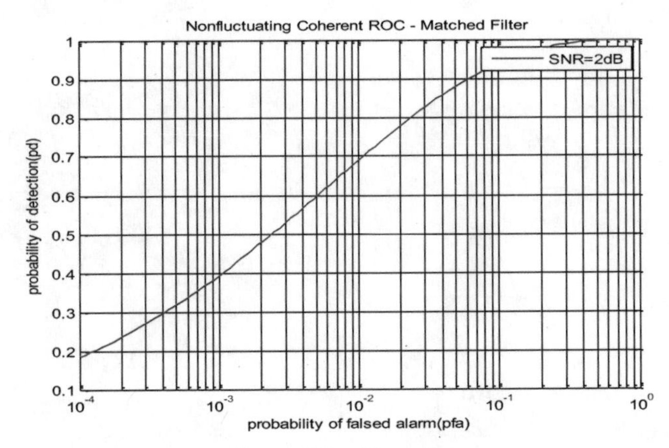

Figure 7.3 Pd vs pfa.

INFERENCE:

The observation from the Figure 7.3 is the presence of a trade off between Pd and Pfa values. For SNR = 4 db, the detection probability approaches 1after Pfa = 0.5.

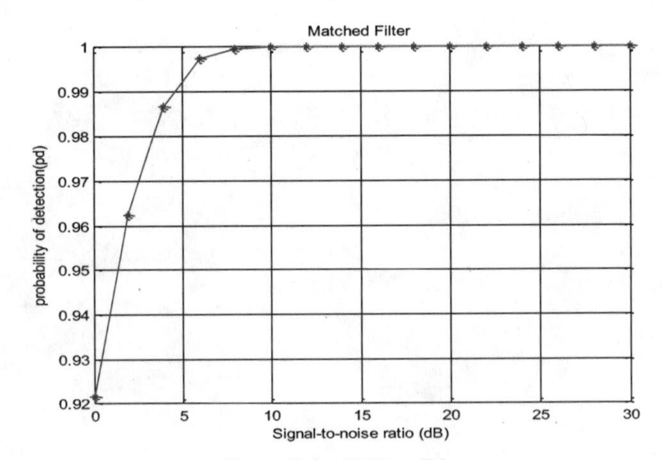

Figure 7.4 SNR vs Pd.

INFERENCE:

The observation from the Figure 7.4 is the presence of a linear increase in the probability of detection for increasing SNR values. The detection probability is high even for low values of SNR. Above 5 db the detection probability reaches 1.

7.3 Cyclostationary Feature Detection

AIM:

To detect the PU (Primary user) signal exploiting the cyclostationary features (spreading code, cyclic prefixes etc). Cyclostationary feature detection is an alternative detection method. Modulated signals are combined with sine wave carriers, pulse trains, repeating spreading, hopping sequences or cyclic prefixes which end up in built in periodicity. These modulated for the reason that their mean and autocorrelation display periodicity feature. Analysis of the spectral correlation function helps detection of such periodicity feature. The spectral correlation function can distinguish between noise energy and modulated signal energy. This arises from the feature of noise as a wide-sense stationary signal and no correlation. On the other hand, modulated signals have the feature of being cyclostationary with special correlation in view of the embedded redundancy of signal periodicity. Therefore, it is possible for a detector of cyclostationary feature top perform better than the energy detector for discrimination against noise. This is because of its robustness to uncertainty seen in noise power. But, there is the computational complexity entailing a long duration for observation.

MATLAB CODE:

```
clc;
clear all;
close all;
r = randi([0 1],100,1);
stem(r,‘DisplayName’,‘r’)
hold on;
l=length(r);
r(l+1)=0;
n=1;
L=1000;
fs=1000;
```

```
T=1/fs;
fc=3/T;
t1 = (0:L);
while n<=l
    t=(n-1):.001:n;
    if r(n)==1
        if r(n+1)==r(n)
            y=(t<=n);
        else
            y=(t<n);
        end
    else
        if r(n+1)==r(n)
            y=(t>n);
        else
            y=(t>=n);
        end
    end
   plot(t,y,'g');
   title('NRZ encoding');
   grid on;
   xlabel('time');
   ylabel('amplitude');
  hold on;
   axis([0 100 -1.5 1.5]);
n=n+1;
end
 b= (2*3.142*fc*t1) +( 3.142*(1-y));
   s = sqrt(2)*cos(b);
   figure,plot(t1,s);
   axis([0 1000 -1.5 1.5]);
title('BPSK modulation');
xlabel('time');
ylabel('amplitude');
```

```
 hold on;
  noise= awgn(s,20,'measured');
  snr_db = 10*log10(s.^2/(noise.^2));
  snr_avg=power(10,snr_db/10);
  rs1=s+noise;
  vr=var(rs1);
 figure,plot(t1,rs1);
 axis([0 100 -1.5 1.5]);
 title('received signal with noise');
 xlabel('time');
 ylabel('amplitude');
 hold on;
 nfft=2^nextpow2(L);
 f1 = fft(rs1,nfft)/L;
 fs1=fs/2*linspace(0,1,nfft/2+1);
figure,plot(fs1,2*abs(f1(1:nfft/2+1)));
title('FFT of received signal');
xlabel('frequency(HZ)');
ylabel('mag|rs1|');
grid on;
hold on;
fc=1:3/T;
s1=cos(2*3.14*fc*T);
l1=length(s1);
sc=s1+c;
l2=length(sc);
cr1=xcorr2(rs1,s1);
figure,plot(cr1);
title('crosscorr of received signal with cosine');
xlabel('frequency(hz)');
ylabel('amplitude');
axis([0 100 -5.5 5.5]);
hold on;
cr2=xcorr2(rs1,sc);
l2=length(cr2);
```

```
figure,plot(cr2);
title('crosscorr of received signal with sine+cosine');
xlabel('frequency(hz)');
ylabel('amplitude');
axis([0 100 -5.5 5.5]);
hold on;
for i=1:500
        if s1(i)>0.5 ||  sc(i)$>$0.5
            display('PU PRESENT');
        else
            display('noise');
        end
end
L=2:2:30;
snr1=2:2:30;
q1=(2*L+1)*(0.5)^2;
Pf=exp(-q1);
del=(2*snr1+1)*vr/(2*L+1);
q2=sqrt(2*snr1/vr);
q3=0.5/del;
Pd=marcumq(q1,q2);
figure;
P=plot(snr1,Pd,'-*');
set(P,'linewidth',2);
title('Pd Vs snr');
xlabel('snr');
ylabel('Probability of detection');
hold on;
grid on;
figure;
P2=semilogx(Pf,Pd,'-*');
set(P2,'linewidth',1.5);
title('Pf Vs Pd');
xlabel('Probability of false alarm');
ylabel('Probability of detection');
hold on;
grid on;
```

RESULTS:

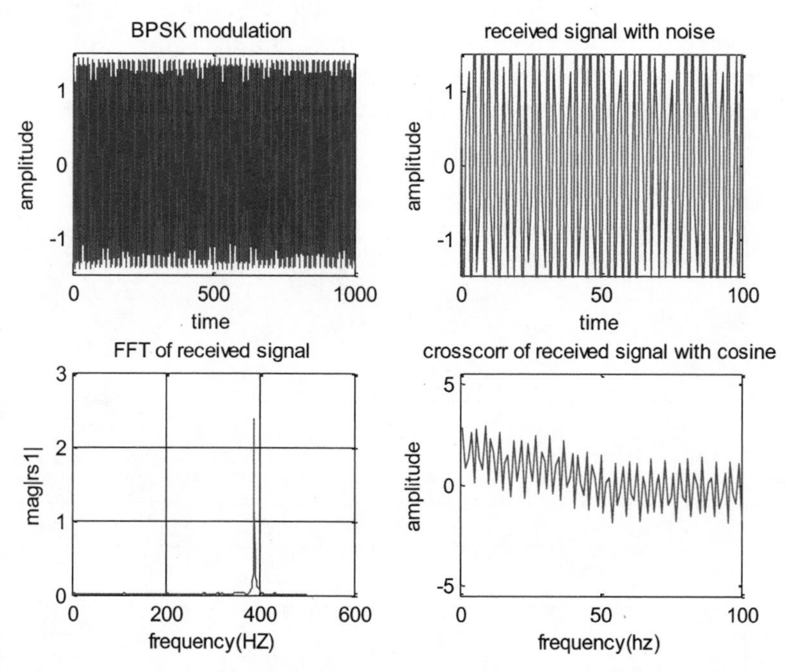

Figure 7.5 a) BPSK modulation b) Received signal with noise c) FFT of received signal d) Cross correlation of received signal with cosine signal.

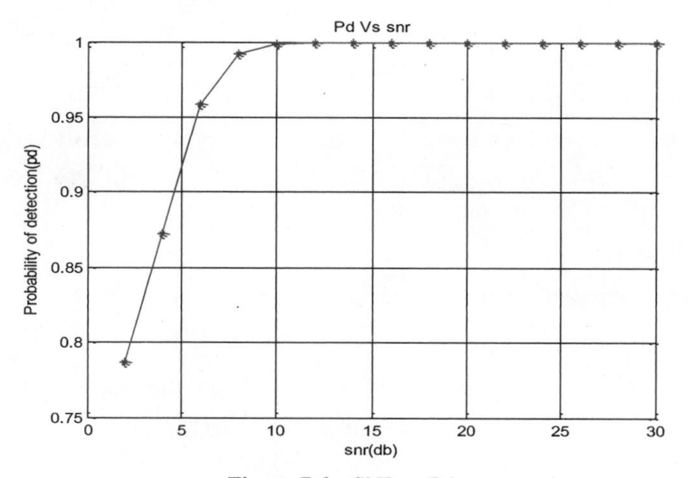

Figure 7.6 SNR vs Pd.

INFERENCE:

The observation from the Figures 7.5 and 7.6 is the presence of a linear increase in the probability of detection for increasing SNR values. For SNR = 2 dB, probability of detection is increased compared to energy and matched filter detection. Finally, the detection probability reaches 1 at SNR = 10 db.

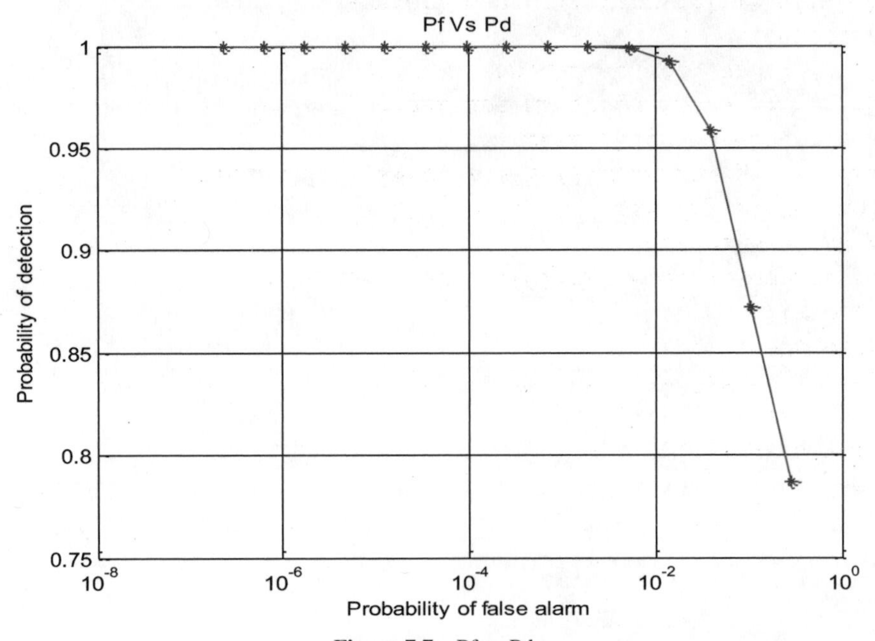

Figure 7.7 Pf vs Pd.

INFERENCE:

The observation from the Figure 7.7 is the presence of a tradeoff between Pf and Pd. For the lower values of Pf up to .001, probability of detection is 1and detection of PU is less for more values of false alarm.

7.4 Co-Operative Spectrum Sensing

AIM:

To enhance the sensing performance by exploiting the spatial diversity in the observations of spatially located CR users. Theoretically, there is greater accuracy seen in cooperative detection among unlicensed users with the possibility of uncertainty in a single user's detection getting minimized.

The features of multipath fading and shadowing effect undermine the performance of primary user detection methods. However, cooperative detection schemes permit mitigation of multipath fading and shadowing effects, thereby improving the detection probability in a heavily shadowed ambience.

MATLAB CODE:

```
clear all;
close all;
clc
u=1000;%time bandwidth factor
N=2*u;%samples
a=2;%path loss exponent
C=2;%constant losses
Crs=10; %Number of cognitive radio users
PdAnd=0;
%----------Pfa------------%
Pf=0:0.01:1;
Pfa=Pf.^2;
%---------signal-----%
t=1:N;
s1 = cos(pi*t);
% s1power=var(s1);
%-------- SNR ----------%
% Snrdb=-15:1:15;
Snrdb=15;
Snreal=power(10,Snrdb/10);%Linear Snr
% while Snrdb<15
for i=1:length(Pfa)
lamda(i)=gammaincinv(1-Pfa(i),u)*2; %thershold
% lamdadB=10*log10(lamda);
%---------Local spectrum sensing---------%
for j=1:Crs %for each node
detect=0;
% d(j)=7+1.1*rand(); %random distanse
d=7:0.1:8;
PL=C*(d(j)^-a); %path loss
for sim=1:10%Monte Carlo Simulation for 100 noise
realisation
%-------------AWGN channel--------------------%
```

```
noise = randn(1,N); \%Noise production with zero mean
and s^2 var
noise_power = mean(noise.^2); %noise average power
amp = sqrt(noise.^2*Snreal);
s1=amp.*s1./abs(s1);
% SNRdB_Sample=10*log10(s1.^2./(noise.^2));
Rec_signal=s1+noise;%received signal
localSNR(j)=mean(abs(s1).^2)*PL/noise_power;%local snr
Pdth(j,i)=marcumq(sqrt(2*localSNR(j)),sqrt(lamda(i )),u);
  %Pd for j node
%Computation of Test statistic for energy detection
Sum=abs(Rec_signal)*PL;
Test(j,sim)=sum(Sum);
if (Test(j,sim)>lamda(i))
detect=detect+1;
end
end %END Monte Carlo
Pdsim(j)=detect/sim; %Pd of simulation for the j-th CRuser
end
PdAND(i)=prod(Pdsim);
PdOR(i)=1-prod(1-Pdsim);
end
PdAND5=(Pdth(5,:)).^5;
Pmd5=1-PdAND5;
PdANDth=(Pdth(Crs,:)).^Crs;
PmdANDth=1-PdANDth; %Probability of miss detection
Pmdsim=1-PdAND;
figure(1);
plot(Pfa,Pmdsim,'r-*',Pfa,PmdANDth,'k-o',Pfa,Pmd5,'g-*');
title('Complementary ROC of Cooperative sensing with AND
      rule under AWGN');
grid on
axis([0.0001,1,0.0001,1]);
xlabel('Probability of False alarm (Pfa)');
ylabel('Probability of Missed Detection (Pmd)');
legend('Simulation','Theory n=10','Theory n=5');
```

RESULTS:

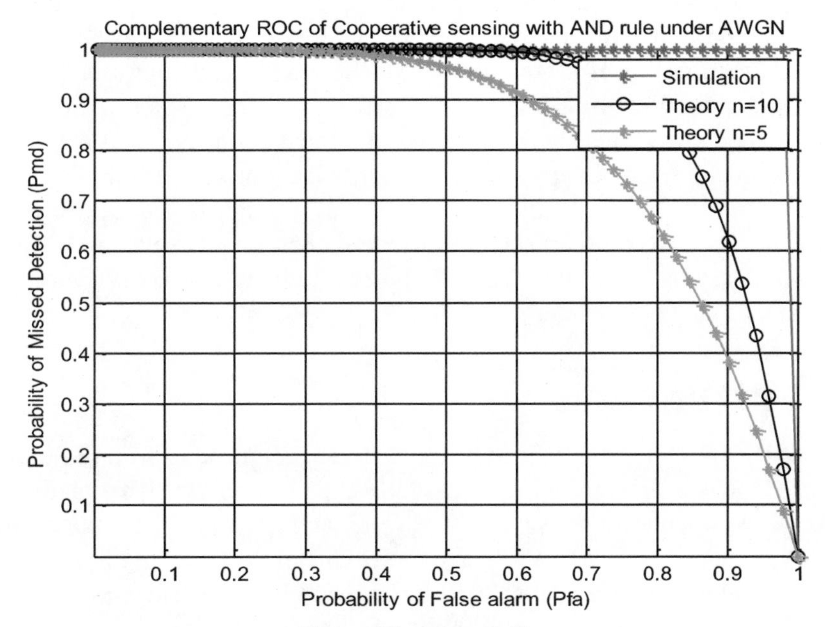

Figure 7.8 Pfa vs Pmd.

INFERENCE:

The observation from the Figure 7.8 is that, observed that, for lower values of Pfa probability of missed detection is 1 and as the value increases probability of missed detection value gradually decreases. As the number of cognitive radio users is increased from 5 to 10, the performance of missed detection is increased but there is a tradeoff between simulation and theory performance. In results of simulation, Pmd value remains 1 up to 0.99 value of Pfa and later approaches 1.

7.5 Introduction & Specifications of USRP

Universal Software Radio Peripheral (USRP) is a range of software-defined radios designed and vended by Ettus Research and its parent company, National Instruments. Developed by a team led by Matt Ettus, the USRP product family is meant to be a comparatively inexpensive hardware platform for software radio, and is broadly used by research labs, universities, and hobbyists. Most USRPs connect to a host computer through a high-speed link, which the host-based software uses for controlling the USRP hardware and transmit/receive data. Some USRP models also integrate the general functionality of a host computer with an embedded processor that enables operation of the USRP device in a stand-alone fashion. Design of USRP family was meant for accessibility, and many of the products are open source hardware. The board schematics for selection of USRP models are available freely for download; control of all USRP products and is done with the open source UHD driver, which is free and open source software. USRPs are commonly used with the GNU Radio software suite to create complex software-defined radio systems.

7.6 USRP B200/B210

The USRP B200 and B210 hardware covers RF frequencies from 70 MHz to 6 GHz, has a Spartan6 FPGA, and USB 3.0 connectivity. This platform makes experimentation possible with a wide range of signals including FM and TV broadcast, cellular, Wi-Fi, and more. One of the USRP B200 features receives and transmits a channel in a bus-powered design. The USRP B210 extends the capabilities of the B200 by offering a total of two receive and two transmit channels, incorporates a larger FPGA, GPIO, and includes an external power supply. Both use an Analog Devices RFIC for delivering a cost-effective RF experimentation platform. It can stream up to 56 MHz of instantaneous bandwidth over a high bandwidth USB 3.0 bus on select USB 3.0 chipsets (with backward compatibly to USB 2.0). Consideration of the B200 and B210 is enabled with USRP Hardware Drive (UHD), users can develop their applications and seamlessly port their designs to high-performance or embedded USRPs such as the USRP X310 or USRP E310. UHD is an open-source, cross-platform driver that has the capability to run on Windows, Linux, and MacOS. It provides a common API, which is used by several software frameworks, such as GNU Radio. With this software support, users can collaborate with a vibrant community of enthusiasts, students, and professionals who have adopted USRP products for their development. As members of this community, users can find assistance for application development, share knowledge to further SDR technology, and contribute their own innovations.

7.7 Experiment-Detection of Spectrum Holes Using USRP

AIM:

To detect spectrum holes in FM Band. Frequency spectrum is the scarcest resource for wireless communication. Cognitive radio can be used for making efficient use of the spectrum. The primary task in cognitive radio is spectrum sensing. It refers to the process of detecting spectrum holes (part of the spectrum which is not utilized). Some methods of spectrum sensing include energy detection, matched filter detection. In this experiment, the spectrum holes in FM Band from 88 to 108 MHz are detected using GNU Radio and USRP kit.

PROCEDURE:

1. Open GNU radio companion.
2. From the block list that appears on the right hand side press CTRL+F to search for the blocks.
3. Click and drag the blocks from the block list to the Canvas (the flow graph construction area).
4. Connect ports by clicking on the chosen port of one block, and then click on the port of the other block. You can delete connections by clicking on the connection's line and pressing the Delete key.
5. The properties of particular block can be changed by double clicking the block and necessary parameters can be changed in the appearing properties window.

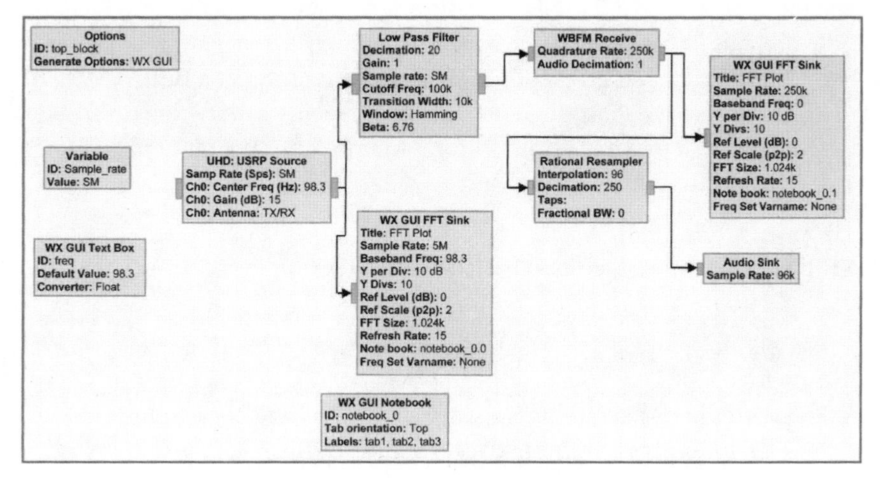

Figure 7.9 Block diagram.

6. Select UHD: USRP source block for implementing in real time so as to interface the hardware with GNU radio.
   The block diagram appears in Figure 7.9.
7. After completing the construction of block run it by using execute option. It is illustrated in Figure 7.10.

RESULTS:

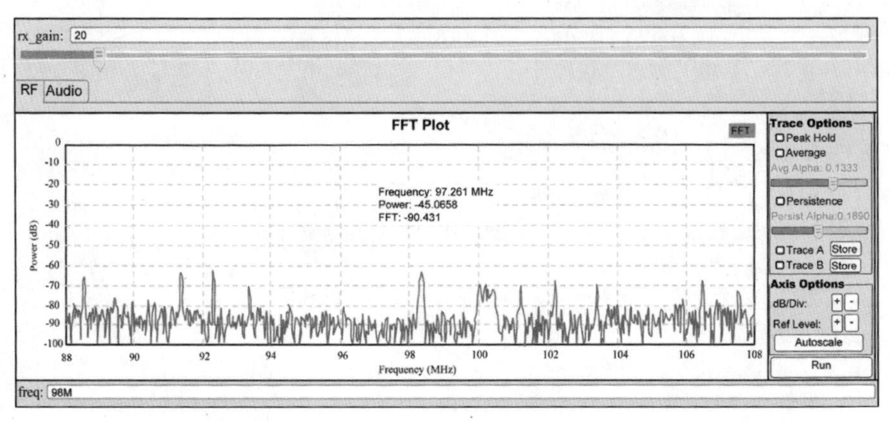

Figure 7.10 FM signal in frequency domain.

7.8 Experiment: Spectrum Sensing Using WARP

AIM:

To see the spectrum of a 2.4 GHz band using WARP. The sensing to be done for idle and busy channel periods over 3.35 seconds of reception at 20 MHz of bandwidth. With over 3 seconds of received waveform at a full 20 MHz of bandwidth, we can see a large extent of channel activity at very fine timescales (25 ns samples). A popular visualization of frequency content across time is known as a spectrogram.

STEPS:

The Figure 7.11 shows the three basic steps to the simple spectrogram in this example:

1. The entire received vector of samples is treated as M sequences of N samples each.
2. The vector is reshaped into a matrix of M rows and N columns.
3. An FFT for each of the M rows is performed. Each FFT is length N.

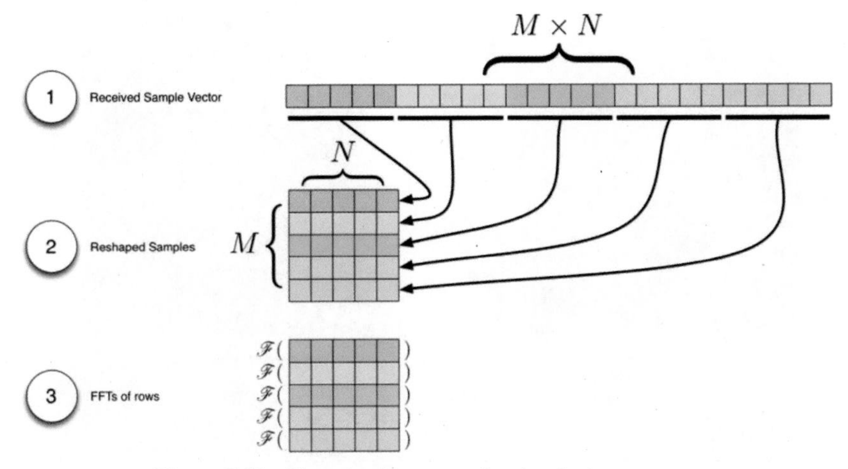

Figure 7.11 Three basic steps to the simple spectrogram.

RESULTS & INFERENCE:

Case1: No Active traffic

For the first case, investigate the channel activity without explicitly introducing any traffic. Take a spectrogram on channel 6, where there are a number of commercial Wi-Fi access points.

The Figure 7.12 shows the raw, unprocessed I and Q samples from the receive buffer as a function of time. Over the 3.35 seconds of reception, it is

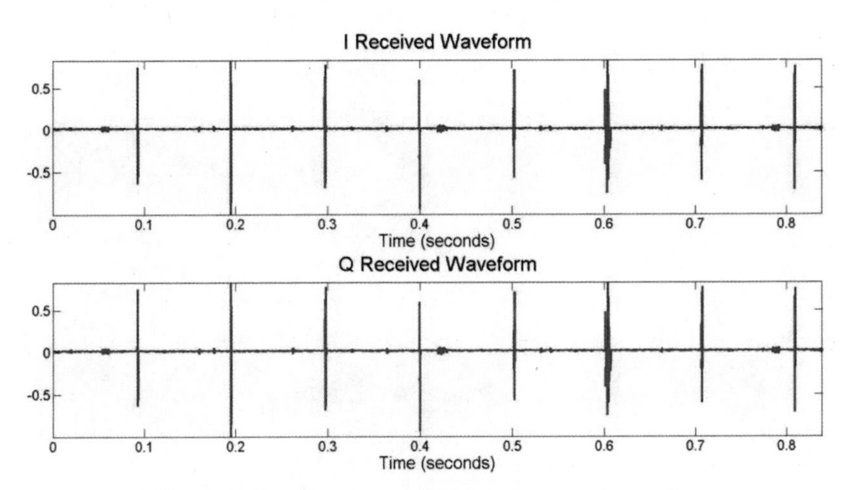

Figure 7.12 Waveforms of I&Q during no active traffic.

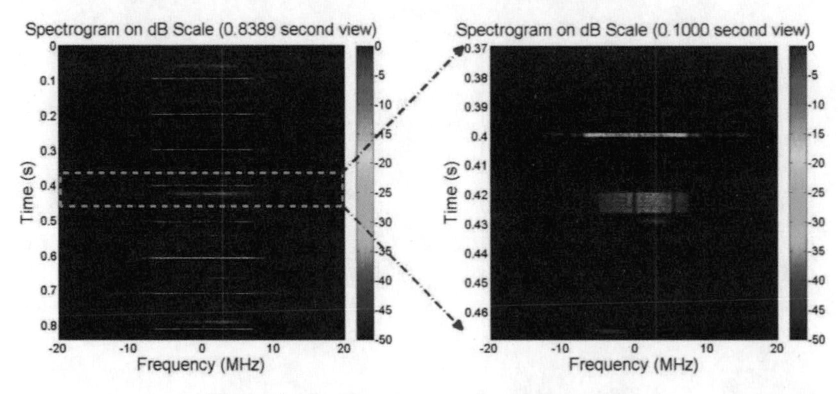

Figure 7.13 Spectrogram of no active traffic.

clear that there is some sort of periodic source of energy on the medium with an interval of 100 ms.

The left subplot of the above Figure 7.13 shows the entire spectrogram of the received waveform. The periodic "bursts" of energy from the time-series plot show up as horizontal lines of energy in the spectrogram. In this view, the periodic energy bursts are seen as approximately 20 MHz wide. At 100 ms intervals, there is energy that is 20 MHz wide. These are 802.11 beacons from our commercial Wi-Fi access point on channel 6.

The right subplot of the Figure 7.13 shows the same data, but zoomed into a small 100 ms slice of the 3.35 s data. You can see that activity is recorded

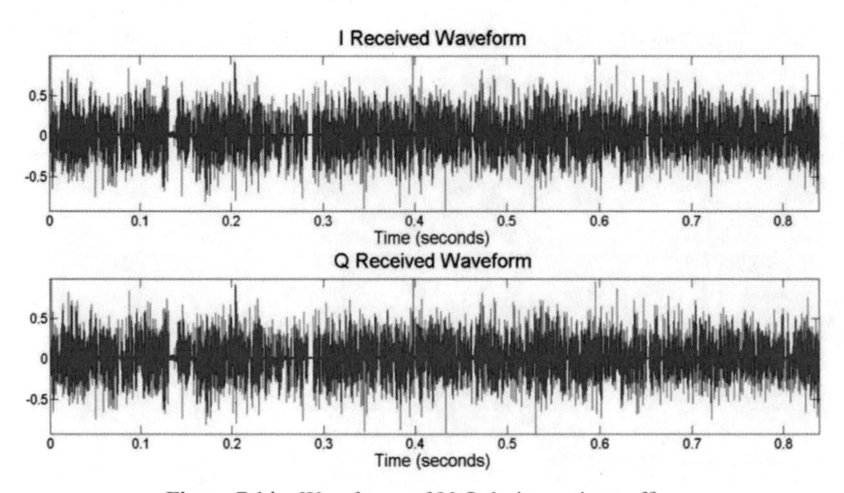

Figure 7.14 Waveforms of I&Q during active traffic.

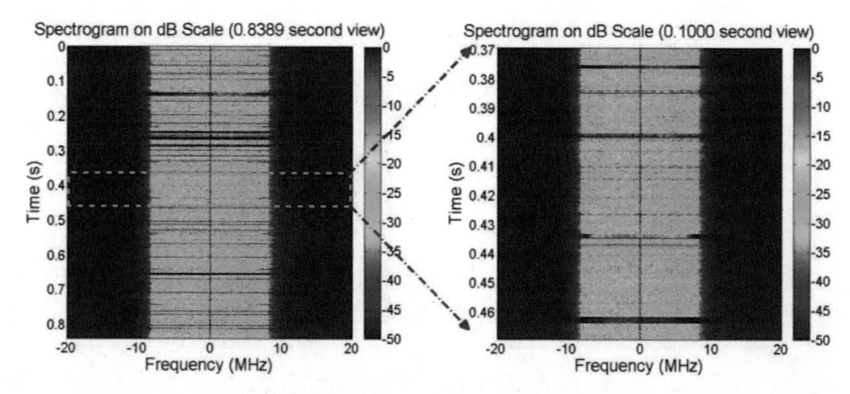

Figure 7.15 Spectrogram of active traffic.

at a much finer granularity than is implied by the left subplot. There is much more data in the spectrogram than can be plotted at reasonable resolutions.

Case2: Active traffic

For the second case, the channel activity when actively was using the RF medium. Specifically, a TCP speed test is ran on a Laptop via a Wi-Fi link with our commercial access point. It is shown in Figure 7.14.

Figure 7.15 shows a marked increase in medium activity in the presence of the non-trivial Wi-Fi traffic. Even in the zoomed-in 100 ms view, many transmissions are seen taking place. It can be zoomed even further and explicitly measure constants defined by the 802.11 standard like the 16 μs SIFS period between a data transmission and its associated ACK.

Bibliography

[1] FCC, Spectrum Policy Task Force, ET Docket 02–135, Nov. 2002.

[2] FCC, Spectrum policy task force report, ET Docket No. 02–380 and No. 04–186, Sep. 2010.

[3] Budiarjo et al., Cognitive Radio Dynamic Access Techniques, Wireless Personal Communications, Springer (2008) 45:293–324.

[4] S. Kumar et al., Cognitive Radio Concept and Challenges in Dynamic Spectrum Access for Future Generation Wireless Communication Systems, Wireless Personal Communication, Springer, 2011(59):525–535.

[5] Hrishikesh Venkataraman, Gabriel-Miro Muntean – Cognitive Radio and its Application for Next Generation Cellular and Wireless Networks? Springer, 2012.

[6] Ekram Hossain et al., Evolution and Future Trends of Research in Cognitive Radio: A Contemporary Survey, Wiley Online Library, Wireless Communications and Mobile Computing, 2013.

[7] Lu et al., Ten years of research in Spectrum Sensing and Sharing in Cognitive Radio, EURASIP Journal on Wireless Communications and Networking 2012, 2012:28.

[8] Yonghong Zeng et al., A Review on Spectrum Sensing for Cognitive Radio: Challenges & Solutions, EURASIP Journal on Advances in Signal Processing, 2010.

[9] T. Wang et al., Analysis of effect of primary user traffic on spectrum sensing performance, Communications and Networking, ChinaCOM, 2009, pp. 1–5.

[10] Jwo-Yuh Wu et al., Energy Detection based Spectrum Sensing with Random Arrival and Departure of Primary User's Signal, IEEE, Globecom Workshops, pp. 380–384, 2013.

[11] Y. Lee et al., Cyclostationarity-Based Detection of Randomly Arriving or Departing Signals, Journal of Applied Research and Technology, vol. 12, no. 6, pp. 1083–1091, December 2014.

[12] Xianzhong Xie et al., Improved Energy Detector with Weights for Primary User Status Changes in Cognitive Radio Networks, Hindawi

Publishing Corporation, International Journal of Distributed Sensor Networks, 2014.

[13] Joseph Font-Sengura et al., GLRT based Spectrum Sensing for Cognitive Radio with prior Information, IEEE Transactions on Communications, vol. 58, no. 7, July 2010.

[14] Daniel W. Bliss et al., MIMO Wireless Communication, Lincoln Laboratory Journal, vol. 15, no. 1, 2005.

[15] David Gesbert et al., From Theory to Practice: An Overview of MIMO Space – Time Coded Wireless Systems, IEEE Journal on Selected Areas in Communications, vol. 21, no. 3, April 2003.

[16] Arogyaswami J. Paulraj, An Overview of MIMO Communications – A Key to Gigabit Wireless, Proceedings of the IEEE, vol. 92, no. 2, February 2004.

[17] R. Karthipan, K. S. Vishvaksenan, R. Kalidoss, R. Sureshbabu, "Uplink capacity enhancement in IEEE 802.22 using modified duplex approach", Wireless personal communication, Springer, Vol. 86, no. 2, pp. 635–656, 2016.

[18] R. Karthipan, K. S. Vishvaksenan, R. Kalidoss, Anjana Krishan, "Performance of turbo coded triply-Polarized MIMO-CDMA System for Downlink Communication", Computers and Electrical Engineering, Elsevier, Vol. 56, pp. 182–192, 2016.

[19] R. Kalidoss, M. A. Bhagyaveni, K. S. Vishvaksenan, "A location based duplex scheme for cost effective rural broadband connectivity using IEEE 802.22 cognitive radio based wireless regional area networks", Fluctuation and noise letters, Vol. 13, no. 4, 2014.

[20] M. A. Bhagyaveni, R. Kalidoss, K. S. Vishaksenan, "Introduction to analog and digital communication", River Publishers, Netherlands, 2016.

[21] R. Kalaiarasan, K. S. Vishvaksenan, R. Kalidoss, "Hybrid spectrum sensing algorithm for cognitive mobile adhoc networks", International journal of advanced engineering technology, Vol. VII, issue 1, pp. 182–186, 2016.

[22] Rajakani, K. Kalidoss. R and Bhagyaveni, M. A., Adaptive duplex technique for reduction of turnaround time in IEEE 802.22. *Global Trends in Computing and Communication Systems*, pp. 503–511, 2012.

[23] Vijayarangan, V., Kalidoss, R. and Sukanesh, R., November. Low crest mapping for PAPR reduction in OFDM systems. *IET International Conference on Wireless, Mobile and Multimedia Networks,* pp. 1–4, 2006.

[24] Vijayarangan, V., Kalidoss, R. and Sukanesh, R., Crest Factor Reduction in Multicarrier Transmission by Low Crest Mapping. *IEEE First*

International Symposium on Pervasive Computing and Applications, pp. 758–763, 2006.

[25] Vishvaksenan, K. S., Mithra, K., Kalidoss, R. and Karthipan, R., Experimental Study on Elliott Wave Theory for Handoff Prediction. *Fluctuation and Noise Letters*, Vol. 15, 04, 2016.

[26] Kalaiarasan, R., Vishvaksenan, K. S. and Kalidoss, R., 2016, March. Performance analysis of Elliot Wave Theory in wireless communication. *International Conference on, Wireless Communications, Signal Processing and Networking (WiSPNET)*, pp. 1864–1868, 2016.

[27] Karthipan, R., Vishvaksenan, K. S., Kalidoss, R. and Krishan, A., Performance of Cognitive Radio based MC-DS-CDMA system for downlink communication, IEEE *International Conference on Wireless Communications, Signal Processing and Networking (WiSPNET)*, pp. 401–404, 2016.

[28] Vishvaksenan, K. S., Kalaiarasan, R., Kalidoss, R. and Karthipan, R., Real Time Experimental Study and Analysis of Elliott Wave Theory in Signal Strength Prediction. *Proceedings of the National Academy of Sciences, India Section A: Physical Sciences*, pp. 1–13.

[29] P. He, L. Zhao, S. Zhou, and Z. S. Niu, "Water-filling: A Geometric Approach and Its Application to Solve Generalized Radio Resource Allocation Problems," *IEEE Trans. Wireless Communications*, vol. 12, pp. 3637–3647, July 2013.

[30] L. Zhang, Y. C. Liang, and Y. Xin, "Joint Beamforming and Power Allocation for Multiple Access Channels in Cognitive Radio Networks," *IEEE J. Select. Areas Communications*, vol. 26, pp. 38–51, Aug. 2008.

[31] P. Wang, X. F. Zhong, L. M. Xiao, S. D. Zhou, and J. Wang, "A General Power Allocation Algorithm for OFDM-Based Cognitive Radio Systems," *IEEE International Conference on Communications Workshops*, pp. 1–5, 2009.

[32] S. C. Yan, P. Y. Ren, and F. S. Lv, "Power Allocation Algorithms for OFDM-Based Cognitive Radio Systems," *IEEE International Conference on Wireless Communications Networking and Mobile Computing(WiCOM)*, pp. 1–4, 2010.

[33] S. Haykin, "Cognitive radio: Brain Empowered Wireless Communications," *IEEEJ. Selet. Areas Communications*, vol. 23, pp. 201–205, Feb. 2005.

[34] I. F. Akyildiz, W. Y. Lee, M. C. Vuran, and S. Mohanty, "Next generation/dynamic spectrum access/cognitive radio wireless networks: A survey," *Computer Networks Journal*, vol. 50, pp. 2127–2159, May 2006.

[35] W. C. Jakes, *Microwave Mobile Communications*. New York, Wiley, 1993.

[36] Y. Nasser, M. des Noes, L. Ros, and G. Jourdain, "Sensitivity of OFDM-CDMA Systems to Carrier Frequency Offset," in *IEEE International Conference on Communications*, vol. 10, pp. 4577–4582, June 2006.

[37] R. Nee and R. Prasad, *OFDM for Wireless Multimedia Communications*. Artech House Publishers, 2000.

[38] L. Hanzo, M. Munster, B. J. Choi, and T. Keller, *OFDM and MC-CDMA for Broadband Multi-User Communications, WLANs and Broadcasting*. John Wiley and Sons Ltd., West Sussex, England, 2003.

[39] S. B. Weinstein and P. M. Ebert, "Data Transmission by Frequency Division Multiplexing Using the Discrete Fourier Transform," *IEEE Transactions on Communications*, vol. 19, pp. 628–634, October 1971.

[40] N. Devroye, P. Mitran, and V. Tarokh, "Achievable Rates in Cognitive Radio Channels," *IEEE Trans. Info. Theory*, vol. 52, pp. 1813–1827, May 2006.

[41] J. Wu, L. Yang, and L. Xu, "Resource Allocation Based on Linear Waterfilling Algorithm in CR Systems," *IEEE International Conference on Wireless Communications, Networking and Mobile Computing (WiCOM)*, pp. 1–4, 2011.

[42] C. H. Chen and C. L. Wang, "An efficient power allocation algorithm for multiuser OFDM-based cognitive radio systems," *in Proc. IEEE Wireless Communications and Networking Conf.*, pp. 1–6, 2010.

[43] J. Jang and K. B. Lee, "Transmit power adaptation for multiuser OFDM systems," *IEEE J. Select. Areas Communications*, vol. 21, pp. 171–178, 2003.

[44] G. Bansal, J. Hossain, and V. K. Bhargava, "Adaptive Power Loading for OFDM-Based Cognitive Radio Systems with Statistical Interference Constraint," *IEEE Trans. Communications*, vol. 10, pp. 2786–2791, July 2011.

[45] A. Goldsmith, S. Jafar, I. Maric, and S. Srinivasa, "Breaking spectrum gridlock with cognitive radios: An information theoretic perspective," *IEEE J. Select. Areas Communications*, vol. 97, pp. 894–914, May 2009.

[46] P. Wang, M. Zhao, X. F. Zhong, L. M. Xiao, S. D. Zhou, and J. Wang, "Power Allocation in OFDM-Based Cognitive Radio Systems," *IEEE Global Telecommunications Conference*, pp. 4065–4065, 2007.

[47] Z. Yonghong and C. Leung, "Resource Allocation in an OFDM-Based Cognitive Radio System," *IEEE Transactions on Communications*, vol. 57, pp. 1928–1931, July 2009.

[48] H. Yaiche, R. R. Mazumdar, and C. Rosenberg, "A game theoretic framework for bandwidth allocation and pricing in broadband networks," *IEEE/ACM Trans. Netw.*, vol. 8, pp. 667–668, Oct. 2000.

[49] F. Wang, M. Krunz, and S. Cui, "Price-Based Spectrum Management in Cognitive Radio Networks," *IEEE Journal of Selected Topics in Signal Processing*, vol. 2, Feb. 2008.

[50] C. Shi, R. Berr, and M. Honig, "Distributed interference pricing for OFDM wireless networks with non-separable utilities," *Information Sciences and Systems*, pp. 755–760, Mar. 2008.

[51] F. F. Digham, "Joint Power and Channel allocation for Cognitive Radios," *IEEE International Conference on Wireless Communications and Networking*, pp. 882–88, Apr. 2008.

[52] A. Jovicic and P. Viswanath, "Cognitive radio: An information-theoretic perspective," *IEEE International Conference on Information Theory*, pp. 2413–2417, July 2006.

[53] W. Yu, G. Ginis, and J. Cioffi, "Distributed Multiuser Power Control for Digital Subscriber Lines," *IEEE J. Select. Areas Commun.*, vol. 20, pp. 1105–1115, June 2002.

[54] X. Wang and Q. Zhu, "Power control for cognitive radio base on game theory," *IEEE International Conference on Wireless Communications, Networking and Mobile Computing*, pp. 1256–1259, Sept. 2007.

[55] D. J. Goodman and N. B. Mandayam, "Power control for wireless data," *IEEE Personal Communications Magazine*, vol. 7, pp. 48–54, Apr. 2000.

[56] S. M. Almalfouh and G. L. Stuber, "A two-step resource allocation algorithm in multicarrier based cognitive radio systems," *IEEE International Conference on Wireless Communications and Networking*, vol. 60, pp. 1699–1713, Apr. 2011.

[57] P. Kaligineedi, G. Bansal, and V. K. Bhargava, "Power Loading Algorithms for OFDM-Based Cognitive Radio Systems with Imperfect Sensing," *IEEE Trans. Wireless Communications*, vol. 11, pp. 4225–4230, Dec. 2012.

[58] G. Bansal, P. Kaligineedi, and V. K. Bhargava, "Joint Sensing and Power Loading Algorithms for OFDM-Based Cognitive Radio Systems," *in Proc. IEEE Wireless Communications and Networking Conf.*, pp. 1–5, Apr. 2010.

[59] G. Bansal, J. Hossain, and V. K. Bhargava, "Optimal and Suboptimal Power Allocation Schemes for OFDM-based Cognitive Radio Systems," *IEEE Trans. Communications*, vol. 7, pp. 4710–4718, Nov. 2008.

[60] D. T. Ngo, C. Tellumbra, and H. H. Nguyen, "Resource allocation for OFDM-based cognitive radio multicast networks with primary user activity consideration," *IEEE Transaction on Vehicular Technology*, vol. 59, pp. 1668–1679, May 2010.

[61] G. Bansal, M. J. Hossain, V. K. Bhargava, and T. Le-Ngoc, "Subcarrier and Power Allocation for OFDMA-Based Cognitive Radio Systems With Joint Overlay and Underlay Spectrum Access Mechanism," *IEEE Trans. Vehicular Tech.*, vol. 62, pp. 1111–1112, Mar. 2013.

[62] Z. Hasan, G. Bansal, E. Hossain, and V. K. Bhargava, "Energy-efficient power allocation in OFDM-based cognitive radio systems: A risk-return model," *IEEE Trans. Wireless Communications*, vol. 8, pp. 6078–6088, Dec. 2009.

[63] G. Bansal, Z. Hasan, J. Hossain, and V. K. Bhargava, "Subcarrier and power adaptation for multiuser OFDM-based cognitive radio systems," *National Conference on Communications(NCC)*, pp. 1–5, Jan. 2010.

[64] T. Weiss, J. Hillenbrand, A. Krohn, and F. K. Jondral, "Mutual interference in OFDM-based spectrum pooling systems," *IEEE Vehicular Technol. Conf.*, vol. 4, pp. 1873–1877, May 2004.

[65] L. Zhang, Y. C. Liang, and Y. Xin, "Joint Beamforming and Power Allocation for Multiple Access Channels in Cognitive Radio Networks," *IEEE Journal on Selected Areas in Communications*, vol. 26, pp. 38–51, 2008.

[66] D. P. Palomar and J. R. Fonollosa, "Practical algorithms for a family of waterfilling solutions," *IEEE Transactions on Signal Processing*, vol. 53, pp. 686–695, 2005.

[67] C. H. Chen and C. L. Wang, "An Efficient Power Allocation Algorithm for Multiuser OFDM-Based Cognitive Radio Systems," *IEEE Wireless Communications and Networking Conference (WCNC)*, pp. 1–6, 2010.

[68] J. Wu, L. X. Yang, and X. Liu, "Subcarrier and Power Allocation in OFDM Based Cognitive Radio Systems," *IEEE International Conference on Intelligent Computation Technology and Automation (ICICTA)*, vol. 2, pp. 728–731, 2011.

[69] P. Zhang, L. X. Yang, and X. Liu, "Subcarrier and Power Allocation in OFDM-Based Cognitive Radio Systems," *International Journal of Computer Network and Information Security*, vol. 1, pp. 24–30, 2010.

[70] Q. L. Qi, A. Minturn, and Y. Q. Yang, "An efficient water-filling algorithm for power allocation in OFDM-Based cognitive radio systems," *IEEE International Conference on Systems and Informatics (ICSAI)*, pp. 2069–2073, 2012.

[71] Q. Qi, A. Minturn, and Y. Yang, "An efficient water-filling algorithm for power allocation in OFDM-based cognitive radio systems," *Systems and Informatics (ICSAI)*, pp. 2069–2073, 2012.

[72] J. Liu, Y. Song, S. Yang, J. Song, and B. Ning, "Optimal power allocation in OFDM-based cognitive radio systems," *Cross Strait Quad-Regional Radio Science and Wireless Technology Conference (CSQRWC)*, vol. 2, pp. 877–880, Sept. 2011.

[73] M. A. McHenry, "NSF Spectrum Occupancy Measurements Projects Summary," *Shared Spectrum Company Report*, Nov. 2005.

[74] S. G. Glistic, *Advanced Wireless Commuincations, Second Edition*. John Wiley Sons, Ltd., 2007.

[75] G. Bansal, J. Hossain, and V. K. Bhargava, "Adaptive Power Loading for OFDM Based Cognitive Radio Systems," *IEEE International Conference on Communications*, pp. 5137–5142, June 2007.

[76] A. Wyglinski, "Effects of bit allocation on non-contiguous multicarrier based cognitive radio transceivers," *IEEE Vehicular Technol. Conf.*, pp. 1–5, Sept. 2006.

[77] P. Cheng, Z. Zhang, H. Huang, and P. Qiu, "A distributed algorithm for optimal resource allocation in cognitive OFDMA systems," *IEEE International Conf. onCommun.*, pp. 4718–4723, May 2008.

Index